W0234770

MOLECULAR
BIOLOGY
INTELLIGENCE
UNIT 8

Minor Histocompatibility Antigens: From the Laboratory to the Clinic

Derry Roopenian, Ph.D.
The Jackson Laboratory
Bar Harbor, Maine, U.S.A.

Elizabeth Simpson, M.A., Vet. M.B., F. Med. Sci.
MRC Clinical Sciences Centre
ICSM, Hammersmith Hospital
London, England, U.K.

CRC Press
Taylor & Francis Group
Boca Raton London New York

CRC Press is an imprint of the
Taylor & Francis Group, an informa business

MINOR HISTOCOMPATIBILITY ANTIGENS: FROM THE LABORATORY TO THE CLINIC

Molecular Biology Intelligence Unit

CRC Press
Taylor & Francis Group
6000 Broken Sound Parkway NW, Suite 300
Boca Raton, FL 33487-2742

First issued in hardback 2019

© 2000 by Taylor & Francis Group, LLC
CRC Press is an imprint of Taylor & Francis Group, an Informa business

No claim to original U.S. Government works

ISBN-13: 978-1-57059-599-8 (hbk)

This book contains information obtained from authentic and highly regarded sources. Reasonable efforts have been made to publish reliable data and information, but the author and publisher cannot assume responsibility for the validity of all materials or the consequences of their use. The authors and publishers have attempted to trace the copyright holders of all material reproduced in this publication and apologize to copyright holders if permission to publish in this form has not been obtained. If any copyright material has not been acknowledged please write and let us know so we may rectify in any future reprint.

Except as permitted under U.S. Copyright Law, no part of this book may be reprinted, reproduced, transmitted, or utilized in any form by any electronic, mechanical, or other means, now known or hereafter invented, including photocopying, microfilming, and recording, or in any information storage or retrieval system, without written permission from the publishers.

For permission to photocopy or use material electronically from this work, please access www.copyright.com (http://www.copyright.com/) or contact the Copyright Clearance Center, Inc. (CCC), 222 Rosewood Drive, Danvers, MA 01923, 978-750-8400. CCC is a not-for-profit organization that provides licenses and registration for a variety of users. For organizations that have been granted a photocopy license by the CCC, a separate system of payment has been arranged.

Trademark Notice: Product or corporate names may be trademarks or registered trademarks, and are used only for identification and explanation without intent to infringe.

Visit the Taylor & Francis Web site at
http://www.taylorandfrancis.com

and the CRC Press Web site at
http://www.crcpress.com

Library of Congress Cataloging-in-Publication Data

Minor Histocompatibility Antigens: From the Laboratory to the Clinic / edited by Derry Roopenian and Elizabeth Simpson.
          p. cm -- (Molecular biology intelligence unit)
     Includes bibliographical references and index.
     ISBN 1-57059-599-2 (alk. paper)
     1. Minor histocompatibility antigens. I. Roopenian, Derry.
II. Series.
   [DNLM: 1. Minor Histocompatibility Antigens. QW 573.5.H6 M666
1999]
QR184.34.M54 2000
616.07'92--dc21
DNLM/DLC
for Library of Congress
                                                    99-31656
                                                    CIP

CONTENTS

EDITORS

Derry Roopenian, Ph.D.
The Jackson Laboratory
Bar Harbor, Maine, U.S.A.
Chapter 1, 4

Elizabeth Simpson, M.A., Vet. M.B., F. Med. Sci.
MRC Clinical Sciences Centre
ICSM, Hammersmith Hospital
London, England, U.K.
Chapter 6

CONTRIBUTORS

Donald W. Bailey, Ph.D.
The Jackson Laboratory
Bar Harbor, Maine, U.S.A.
Chapter 2

Patrick G. Beatty, M.D., Ph.D.
Division of Hematology/Oncology
Department of Medicine
University of Utah School of Medicine
Salt Lake City, Utah, U.S.A.
Chapter 9

Frederike Bemelman, M.D.
Sir William Dunn School of Pathology
South Parks Road
Oxford, England, U.K.
Chapter 14

Marc A. Berger, Ph.D.
Kimmel Cancer Institute
Jefferson Medical College
Philadelphia, Pennsylvania, U.S.A.
Chapter 12

Geoffrey W. Butcher, Ph.D.
Laboratory of Immunogenetics
Cellular Immunology Programme
The Babraham Institute
Cambridge, England, U.K.
Chapter 5

Phillip Chandler, Ph.D.
Transplantation Biology Group
MRC Clinical Sciences Centre
ICSM, Hammersmith Hospital
London, England, U.K.
Chapter 6

Gregory Christianson, B.S.
The Jackson Laboratory
Bar Harbor, Maine, U.S.A.
Chapter 4

Stephen Cobbold, Ph.D.
Sir William Dunn School of Pathology
South Parks Road
Oxford, England, U.K.
Chapter 14

Thea M. Friedman, Ph.D.
Kimmel Cancer Institute
Jefferson Medical College
Philadelphia, Pennsylvania, U.S.A.
Chapter 12

David Ginsburg, M.D.
Departments of Internal Medicine
  and Human Genetics and the
  Howard Hughes Medical Institute
University of Michigan
Ann Arbor, Michigan, U.S.A.
Chapter 11

Els Goulmy, Ph.D.
Department of Immunohaematology
  and Blood Bank
Leiden University Medical Center
Leiden, The Netherlands
Chapter 9

James C. Jenkin, B.A.
Department of Medicine
University of Utah Health Sciences
  Center
Salt Lake City, Utah, U.S.A.
Chapter 9

Sean L. Johnston, M.D., Ph.D.
Departments of Surgery
  and Immunology
The Mayo Foundation
Rochester, Minnesota, U.S.A.
Chapter 13

Robert Korngold, Ph.D.
Kimmel Cancer Institute
Jefferson Medical College
Philadelphia, Pennsylvania, U.S.A.
Chapter 12

Paul J. Martin, M.D.
Fred Hutchinson Cancer Research
  Center
Seattle, Washington, U.S.A.
Chapter 10

Lisa M. Mendoza, Ph.D.
Division of Immunology
Department of Molecular and Cell
  Biology
University of California
Berkeley, California, U.S.A.
Chapter 4

Thalachallour Mohanakumar, M.D.
Department of Surgery
Washington University School
  of Medicine
St. Louis, Missouri, U.S.A.
Chapter 15

Bashoo Naziruddin, M.D.
Department of Surgery
Washington University School
  of Medicine
St. Louis, Missouri, U.S.A.
Chapter 14

William C. Nichols, Ph.D.
Departments of Internal Medicine,
  Human Genetics and the Howard
  Hughes Medical Institute
University of Michigan
Ann Arbor, Michigan, U.S.A.
Chapter 11

Pedro Paz, Ph.D.
Division of Immunology
Department of Molecular and Cell
  Biology
University of California
Berkeley, California, U.S.A.
Chapter 4

Claude Perreault, M.D.
Department of Medicine
University of Montreal
Guy-Bernier Research Center
Maisonneuve-Rosemont Hospital
Montreal, Quebec, Canada
Chapter 8

Stéphane Pion, M.Sc.
Department of Medicine
University of Montreal
Guy Bernier Research Center
Maisonneuve-Rosemont Hospital
Montreal, Quebec, Canada
Chapter 8

Nancy J. Poindexter, M.D.
Department of Surgery
Washington University School
  of Medicine
St. Louis, Missouri, U.S.A.
Chapter 15

Hans-Georg Rammensee, Ph.D.
University of Tübingen
Institute for Cell Biology
Department of Immunology
Tübingen, Germany
Chapter 3

Denis C. Roy, M.D.
Department of Medicine
University of Montreal
Guy Bernier Research Center
Maisonneuve-Rosemont Hospital
Montreal, Quebec, Canada
Chapter 8

Diane Scott, Ph.D.
Transplantation Biology Group
MRC Clinical Sciences Centre
ICSM, Hammersmith Hospital
London, England, U.K.
Chapter 6

Nilabh Shastri, Ph.D.
Division of Immunology
Department of Molecular and Cell
    Biology
University of California
Berkeley, California, U.S.A.
Chapter 4

Craig R. Smith, M.D.
Department of Surgery
Washington University School
    of Medicine
St. Louis, Missouri, U.S.A.
Chapter 15

Debbie Statton, Ph.D.
Kimmel Cancer Institute
Jefferson Medical College
Philadelphia, Pennsylvania, U.S.A.
Chapter 12

David Steinmuller, Ph.D.
Department of Microbiology
Montana State University
Bozeman, Montana, U.S.A.
Chapter 15

Pierre van der Bruggen, Ph.D.
Ludwig Institute for Cancer Research
Brussels, Belgium
Chapter 16

Herman Waldmann, M.D.
Sir William Dunn School of Pathology
South Parks Road
Oxford, England, U.K.
Chapter 14

Peter Wettstein, Ph.D.
Departments of Surgery
    and Immunology
Mayo Foundation
Rochester, Minnesota, U.S.A.
Chapter 13

Matt Wise, M.D., Ph.D.
Sir William Dunn School of Pathology
South Parks Road
Oxford, England, U.K
Chapter 14

Lesley L. Young, Ph.D.
Division of Immunology
Department of Pathology
University of Cambridge
Cambridge, England, U.K.
Chapter 5

Aamir Zuberi, Ph.D.
Pennington Biomedical Research
    Center
Baton Rouge, Louisiana, U.S.A.
Chapter 4

FOREWORD

This is the second volume to focus on minor histocompatibility (H) antigens. It was preceded by a best-selling novel from Robin Cook entitled "Chromosome 6".[1] In this novel, a molecular biologist anti-hero developed a transgenic technique to transpose segments of human chromosomes encoding minor H antigens together with the segment containing the major histocompatibility complex (MHC) into *bonobo* chimpanzees. For the right price, an individual could thus order the creation of a chimp transgenic for his *H* genes. The transgenic *bonabo* would then be the perfectly matched organ donor should the need arise. Unfortunately, the chimps turned out to be uncomfortably human....

The purpose of our book is to provide an easily readable book providing an up-to-date understanding of minor H antigens and the genes that encode them from both experimental and clinical perspectives. The reader will be better equipped to judge whether the anti-hero's heavy-handed and apparently lucrative strategy could succeed. More seriously, a scientific text focused on minor H antigens is particularly timely in the light of recent advances of knowledge. While the MHC is the strongest single barrier to tissue transplantation, mouse studies at the beginning of the 20th century made it apparent that many additional *H* loci were involved. The antigens encoded by minor *H* loci have proven to be a barrier to transplantation in humans as well, especially in the case of bone marrow transplantation (see Chapter 9 by Goulmy). With the widespread application of allogeneic tissue and bone marrow transplantation as a treatment for an array of life threatening disorders, there is a pressing need for experimentalists and clinicians to understand minor H antigens, the problems they pose, and the therapeutic opportunities they present.

Progress is being made on many fronts. Newer biochemical and DNA cloning methods have extracted molecular explanations finally allowing one to ascribe a rational molecular basis to minor H antigens both in rodents and humans. This understanding also paves the way to apply new genomic analyses to the problem of tissue transplantation. Several chapters (e.g., those by Mendoza et al and Goulmy) thus focus on molecular characterization, while others (Martin, Ginsburg and Nichols, Beatty and Jenkin) approach allogeneic bone marrow transplantation as a genetic problem amenable to genomic analysis. But much of the richness of minor antigens lies in their history; the older literature often holds key insights into contemporary problems and the chapter by Bailey is an excellent example of this. Chapters by Simpson et al and Roopenian relate the historical data to advances in understanding autosomal and male-specific (HY) minor H antigens, respectively, culminating in their molecular definitions. Both Steinmuller and Mohanakumar et al discuss older and newer concepts related to tissue specific antigens that result in barriers to organ rejection, and the chapter by Wise et al describes groundbreaking work of their group in infectious tolerance, showing that minor H antigens can be exploited to induce tolerance to additional alloantigens.

Major breakthroughs in the understanding of basic features of immune recognition have emerged from the study of minor antigens. Rammensee recounts how his efforts toward defining minor antigens led to the elucidation of peptide presentation by MHC molecules. Butcher and Young describe the key insights provided by analyses of the first minor H antigens to be identified, the mitochondrial minor H peptides, how they are processed, presented, and elicit T-cell responses. Perreault et al, Wettstein and Johnson, and Korngold et al use the minor H antigen system to approach the important problem of immunodominance in graft versus host disease, but the principles apply to many complex antigenic situations. T cells are the effectors of graft rejection, and Wettstein and Korngold provide new insights into the recognition of minor H antigens from the perspective of the T-cell receptors used. Finally, van der Bruggen recounts how the analysis of tumor antigens has led to intriguing insights into how genes altered or activated as a consequence of neoplastic transformation can be exploited for tumor vaccines and their similarity to minor H antigens.

We thank the contributing authors for their excellent chapters and their cooperation. This volume was conceived as a consequence of the First International Symposium on Minor Histocompatibility Antigens held in Bar Harbor, Maine in the fall of 1997, and includes many advances made over the course of 1998.

Derry Roopenian
Elizabeth Simpson

1. Cook, R. Chromosome 6. New York: Putnam Publishing Group, 1997.

This book is dedicated to the memory of George D. Snell, Geneticist, Ethicist, and Gardener.

CHAPTER 1

Lessons from *H3*, A Model Autosomal Mouse Minor Histocompatibility Locus

Derry Roopenian

Histoincompatibility was first detected in mice at the beginning of the 20th century. The great majority of the loci responsible for this complex genetic trait are located on the autosomal chromosomes. The ability to customize the mouse genome by the production of congenic stocks of mice that isolate individual histocompatibility (H) loci has enormously simplified the analysis of the trait of histoincompatibility. Recent advances in molecular genetics are making it possible to define the minor *H* genes captured in congenic strains. Given the genetic and biological similarity of mice and humans, the molecular characterization of mouse *H* loci increases the value of the mouse model in developing approaches to improve the success of tissue transplantation between humans.

Minor H-Congenic Mouse Strains

A minor H locus is an operational definition first coined by George Snell to describe the genes responsible for rejection when tissues were transplanted between genetically dissimilar mice. Knowing that many genes were involved, George Snell utilized mathematical formulae developed by the noted, eclectic geneticist, J.B.S. Haldane, to develop breeding schemes aimed at producing mice that differed only at a single *H* locus.[1] The scheme was to cross mice from a "donor" strain, such as 129, to a "recipient" strain, such as C57BL/10 (commonly abbreviated to B10). F1 progeny were backcrossed to B10 and then intercrossed. The intercross progeny were then tested to determine whether they resisted the growth of transplanted B10 tumor cells. Most of the mice died because they were not so fortunate as to inherit from 129 an *H* locus that was capable of preventing the growth of the tumor. However, mice that were homozygous for at least one *H* locus allelically variant from B10 survived and were used as founder mice for additional backcross-intercross cycles. In this way the 129 genome was diluted out except for a donor strain-derived *H* locus sufficient to provide resistance to the growth of the transplanted tumor. Through this prodigious effort, he produced a wide assortment of congenic-resistant mouse strains (see ref. 3).[2] With his colleague Ralph Graff, Snell subsequently showed that many of the congenic mice also rejected skin grafts from their partner strains.

Snell's classical experiments with Peter Gorer led to the realization that disparity at a particular *H* locus conferred resistance to the growth of transplanted tumors particularly efficiently.[4] This locus localized with Gorer's serologically defined antigen II, and thus was referred to as *H2*, and later, the major histocompatibility complex (MHC). In comparison with *H2*, the remaining "minor" *H* loci exhibited a weaker and more variable rejection

Minor Histocompatibility Antigens: From the Laboratory to the Clinic, edited by Derry Roopenian and Elizabeth Simpson. ©2000 Landes Bioscience.

phenotype in that they often required preimmunizations to prevent tumor growth.[5] Consecutive H numbers were assigned to each remaining locus, e.g., *H-1, H-3, H-4.* (Note that a recent nomenclature change has removed all dashes from gene symbols, H-1 is now *H1,* etc.). It was possible to establish "linkage" for some of the *H* loci because their respective congenic donor segments carried visually detectable phenotypic markers. For example, mice congenic for H*1* are albino while the recipient strain, B10, is black; thus H*1* is linked to the albino locus (c). *H2* proved to be linked to the fused locus (*fu*), *H3* to agouti (a), and *H4* to pink-eyed dilution (p).[3]

Bailey produced many additional congenic strains by a more straightforward and efficient manner[6,7] (See Chapter 2). C57BL/6 (B6) mice were crossed to BALB/c mice, and the F1 progeny were then repeatedly backcrossed to B6. In each backcross generation, tail skins from the individual backcross mice were transplanted onto normal B6 mice. If the backcross mouse inherited an *H* locus from BALB/c, its skin was rejected. These mice were selected for breeding in subsequent backcross generations, resulting in an encyclopedic array of congenic strains derived from the same genetic background, all of which were based on the ability of B6 mice to reject a skin graft from BALB/c-derived donor segments.

Largely due to the production of congenic strains, approximately 60 mouse minor *H* loci have been mapped to date (Fig. 1.1). The great majority of the loci have been mapped through the analysis of congenic of mouse strains that carry heterologous donor segments on the B10 or B6 recipient genetic background. Bailey (Chapter 2) has predicted that the number of mapped minor *H* genes is a gross underestimate, and that in all likelihood, there are hundreds more.[8,9] As suggested by Johnson,[10] and discussed below, it is thus likely that some minor H antigen congenic segments contain more that one *H* gene.

The *H3* Conundrum

Studies of the *H3* complex are particularly instructive towards understanding how genes within a congenic segment contribute to tissue rejection. The *H3* "locus" was discovered by Snell and coworkers as a consequence of construction of the congenic strain B10.LP. It carries a congenic donor segment derived from the LP strain on a C57BL/10 background.[2] In addition to *H3,* B10.LP mice inherited the dominant white-belly agouti allele (*A^w*) from LP, while B10 mice carried the non-agouti (*a*) allele and therefore are black. Thus linkage of *H3* to agouti, now known to map to mouse chromosome 2, was established.

Well before the advent of molecular genetic techniques, Snell and his collaborators' masterful exploitation of the fortuitous linkage of *H3* to agouti and other visually detectable linked loci greatly facilitated the further analysis of *H3.* From crosses between B10.LP and B10, they produced strains that carried a recombination within the *H3* congenic segment. Mice of one recombinant strain, B10.LP-a, was black, indicating that it had inherited the B10 non-agouti allele. Another was white-belly agouti indicating that it had inherited LP's white-belly agouti allele. F1 complementation (described in ref. 3) showed that both recombinant strains resisted the growth of a B10-derived tumor, suggesting that the two strains rejected the tumor because of different *H* genes. These results proved that there were a minimum of two *H* genes within the *H3* congenic segment:[11,12] the one mapping farthest from agouti retained the designation *H3,* and the other, designated *H13,* mapped close to agouti. They also were able to produce additional congenic strains that carried *H3* segment from a different donor. The now defunct strain, UW, was a chromosome 2 linkage stock, meaning that it was bred to carry three chromosome 2 phenotypic markers, *A^t* (tan belly agouti), wellhaarig (*we*) (causes curly hair trait), and undulated (*un*) (causes a disfigured tail), all within a congenic chromosomal segment delimiting *H3.* The congenic strain B10.UW was produced, which carried all three of these genetic markers in addition to rejecting B10 tumors and skin. Guided by the three visual markers, intra-*H3* recombinant mouse strains

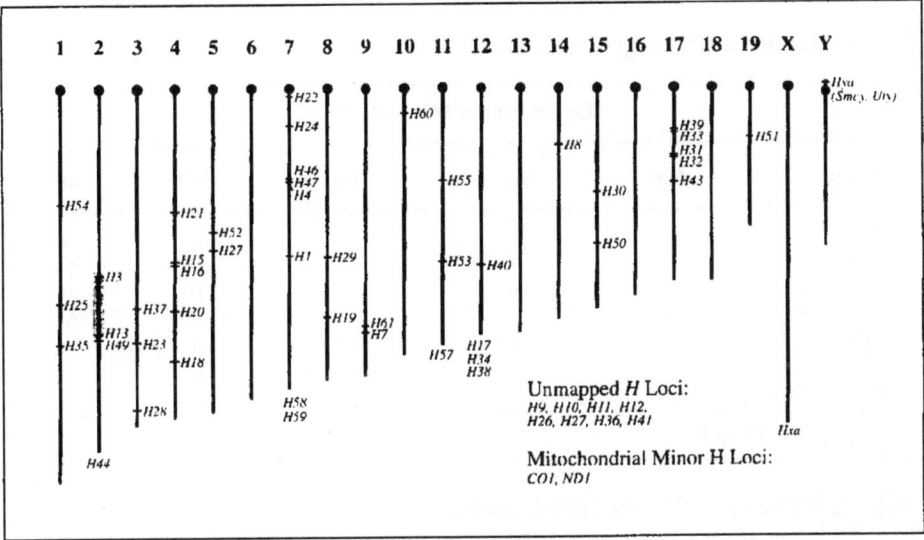

Fig. 1.1. Chromosomal position of mouse minor *H* loci. Adapted from the Mouse Genetic Database (http://www.informatics.jax.org). *H* loci mapping to an unknown position on the indicated chromosomes are also indicated. The blocked region of Chromosome 2 positions the greater *H3* complex.

were produced that further subdivided what was now regarded to be a gene complex (Table 1.1).[11,12]

Using these strains, it was possible to map *H3* close to *we*, and to show that *un* mapped between *H3* and *H13*. Snell noted two curious phenomena illustrated in Figure 1.2. The first was that disparity at the genetic region surrounding *un* seemed to accelerate rejection of grafts that differed at H3.[11,12] The second was that haplotype of *H3* seemed to have a profound effect on the pace of rejection of H13-incompatible skin and bone marrow transplants.[11-13] Rejection of H13-incompatible grafts was strong when host and donor were H3[a], but practically nonexistent when host and donor were H3[b]. An Immune Response (Ir) gene mapping near un and controls both H3 and H13 rejection was among the speculations regarding as to how this result might be interpreted[1].[15,19,20]

---

[1] The issue of an *Ir* gene between *H3* and *H13* is made even more interesting by identification of other *Ir* genes that have been mapped to *H3*. Gasser reported that the antibody response to an erythrocyte antigen, Ea1, mapped to a site between *H3* and *H13*.[14,15] He referred to this locus as Immune Response 2 (*Ir2*). This gene was unusual compared with *H2*-linked *Ir* genes in that responsiveness proved to be a recessive trait. Several other studies have mapped immune response traits near the *H3* complex. This includes CTL responses to HY antigens in the context of H2[d] [16] and H2[b] [17] and the T-cell response to soluble hen egg lysozyme.[18] However, we have been unable to confirm the control of the response to hen egg lysozyme and the CTL response to H2[b]-restricted HY antigens, and the mapping data linking the CTL responses to H2[d] restricted HY antigens to H3 are weak.[16] More recently, a locus that influences autoimmune diabetogenesis in NOD mice also has been mapped to the *H3* region.[21]

Table 1.1. Snell's *H3* congenic strains

	The Greater *H3* Complex[1]				
Strains	***H3***	***we***	***un***	***H13***	***a***
B10	*a*	+	+	*a*	*a*
B10.LP	*b*	+	+	*b*	*A^w*
B10.LP-*a*	*a*	+	+	*b*	*a*
B10.LP-*A^w*	*a*	+	+	*b*	*A^w*
B10.UW	*b*	*we*	*un*	*a*	*A^l*
B10.UW-a	*b*	*we*	*un*	*a*	*a*
B10-*H3^b we*	*b*	*we*	+	*a*	*a*

[1] Based on refs. 11 and 12.

The Advent of In Vitro Techniques

With the advent of in vitro analytic techniques, in particular the cell-mediated lympholysis (CML) assay, it became apparent cytotoxic T-cells were commonly generated following in vivo immunization and restimulation in mixed leukocyte culture (MLC) with minor H antigen-disparate cells.[22,23] This advance made it feasible to dissect the mechanisms of immune recognition and the immunogenetics of congenic strains. In due course CTLs were established as important effectors of graft rejection.[24] Thus, while minor H antigens previously were defined as non-MHC loci that caused tissue rejection in vivo, the elicitation of CTLs became an in vitro correlate for minor H antigens.

My association with minor H antigens began late in the 1970s, while I was a technician in the laboratory of Robert Click, who was more of a mentor than a boss. We had succeeded in generating CTL responses to H3 antigens after immunizing B10.LP-*a* mice with B10 cells. I found it curious that CTL responses generated after immunization against H3 were restricted by both H2K^b and H2D^b, and in my naïveté proposed that the dual restriction might be explained by the existence of two genes, one of which encoded a minor H antigen only presented by H2K^b and the other encoded an antigen only presented by H2D^b. Although skeptical, Click allowed me to test this hypothesis, which ultimately proved true.[25-27] For better or for worse, this early success whet my interest in minor H antigens.

Several years later, after assuming a position at The Jackson Laboratory, I returned to the genetic analysis of *H3* after realizing that not only CTLs but also CD4$^+$ helper T-cells played an integral role in responses against H3 antigens.[28,29] Our prediction was that the *H3* congenic segment must encode epitopes that not only stimulate CD8$^+$ CTLs but also CD4$^+$ helper T cells. We established cloned MHC class II-restricted CD4$^+$ helper T-cells and both class I K^b and D^b-restricted CD8$^+$ T-cell lines, all of which were specifically reactive with B10-derived H3^a antigens. We used them to analyze the genetics of H3 by examining the proliferative response of the cloned T-cell lines mounted in response to spleen cells of F$_2$ mice from crosses between B10.UW and B10 mice. The gene(s) encoding the CTL-defined antigens co-segregated; however, in rare cases recombinants were detected that separated the CTL-defined genes from the gene encoding the helper cell-defined antigen(s)$_2$. Thus the

$_2$Our operational definition for a minor H locus whose antigen is only presented by class I molecules is a type I locus, while a locus whose antigen is only presented by class II molecules is a type II locus.

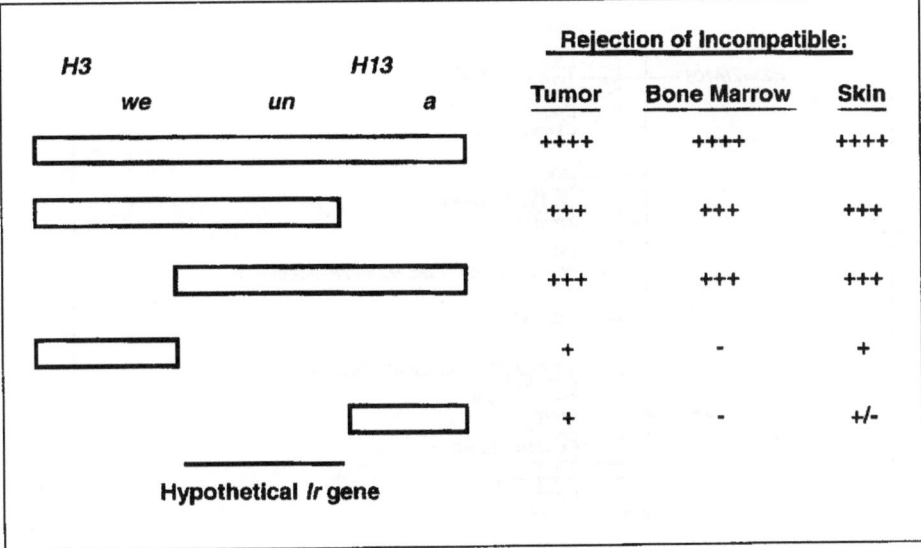

Fig. 1.2. The *H3* conundrum. Boxed regions show the extent of genetic dissimilarity between combinations of the congenic strains described in Table 1. Intensity of rejection is indicated in scale of – (none) to ++++ (intense). Tumor and bone marrow survival of immunized mice from refs. 11,12. Skin graft survival data from refs. 13,14,19.

genes encoding CTL determinants (referred to as type I genes) and helper determinants (referred to as type II genes) were proven to be encoded by separate genes lying within the *H3* complex.

Molecular Genetic Analysis of the Greater H3 Complex

The advent of molecular biological techniques made it possible to approach the molecular genetics of the greater *H3* complex. We produced a high resolution genetic mapping of much of *H3* [27,30-32] onto which we integrated the above-mentioned type I and type II loci (Fig. 1.3). This map proved our earlier contention that there were two genetically distinct, but very closely distinct type I loci. As discussed below, one of them is β_2 microglobulin ($\beta 2M$), which encodes the light chain of class I proteins. We named the other *H3a. B2M* maps distal to *H3a*, and both map proximal to *we*. These type I genes mapped several centiMorgans proximal to the type II gene, which we designated *H3b. H3b* mapped less than 1 centiMorgan proximal of *un*, while *H13* mapped very close to agouti. This map made it clear that the greater *H3* complex encompasses many genes, punctuated with rare genes that display the properties that qualify them as minor *H* genes. Moreover, the map provides a starting point for developing positional cloning approaches to characterize the genes responsible for the functional attributes discussed in the preceding paragraphs.

β2M, an Exceptional Minor H Gene

β_2 microglobulin (β_2M) comprises the light chain of class I molecules. Rammensee and Klein,[33] followed by Kurtz and collaborators,[34] provided the first evidence suggesting that the K^b-restricted activity elicited in responses between *H3* congenic strains was determined by allelic variation of β_2M. Experiments of Pérarnau and collaborators suggested that the mechanism by which the polymorphism affects recognition was that the amino acid

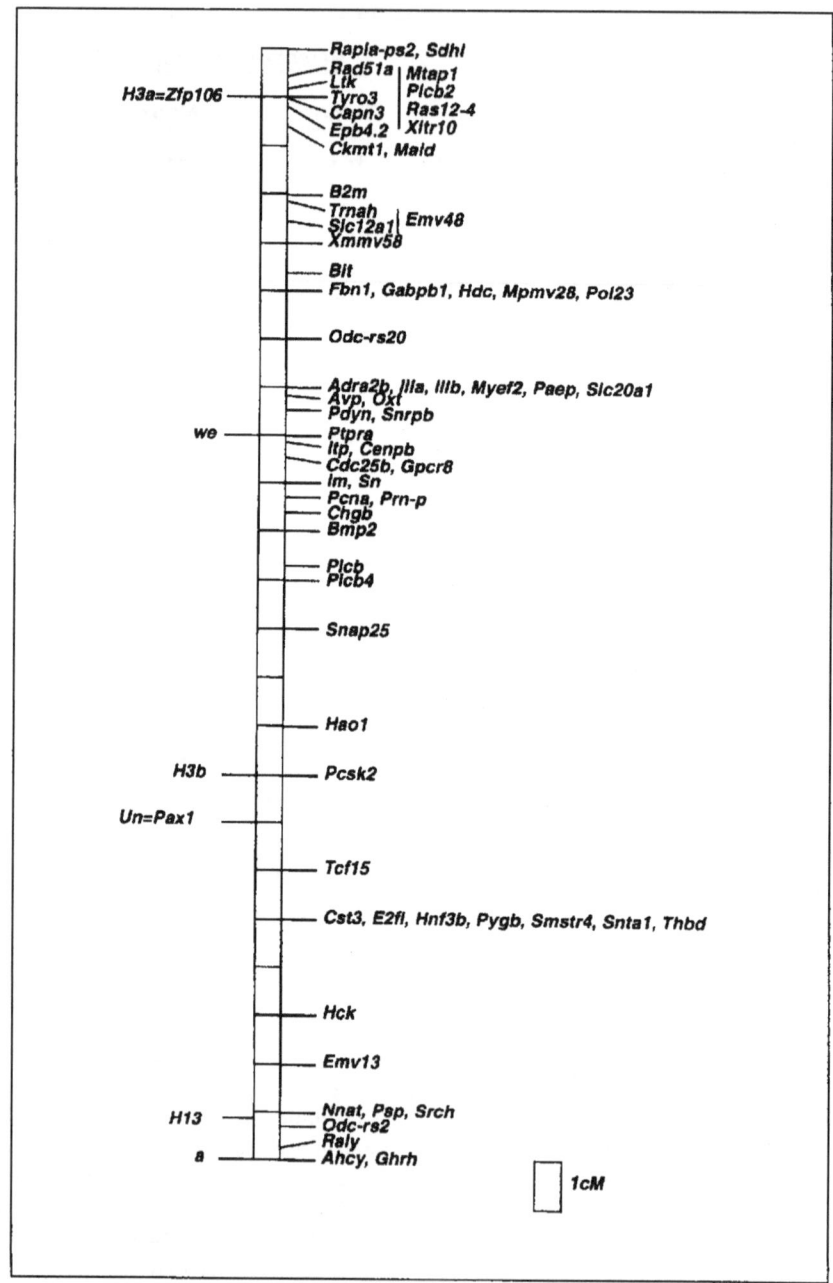

Fig. 1.3. Molecular genetic organization of the greater *H3* complex. Adapted from Mouse Genetic Database (http://www.informatics.jax.org) and refs. 26,29. Order of genes to the right of the vertical line indicates order of loci unknown.

change had a subtle effect on the spectrum of peptides that bind to class I molecules.[35] β_2M allele *a* mice therefore are not tolerant to novel peptides presented by the K^b-β_2M^b heterodimer. The physicochemical basis of β_2M "restriction" is not understood since β_2M *a* and *b* alleles differ by only a single amino acid at position 85 that does not directly contact the α-helices or the β-pleated sheet of the class I protein. Transgenic studies suggest that allelic differences in β_2M require a helper stimulus in order to reject skin.[36] No known allelic variation is found in humans, so β_2M is a tissue transplantation barrier unlikely to act as a minor *H* locus in humans.

H3a, a Prototypic Type I Minor H Gene

H3a is the dominant antigen against which CTL activity is detected in responses between *H3*-congenic mice. To identify the *H3a* gene, Aamir Zuberi, a talented postdoctoral fellow, undertook the heroic task of a positionally cloning this gene. Exploiting the close linkage of *H3a* to *B2M*, he developed a physical map comprised of overlapping yeast artificial chromosomes (YACs) containing DNA from B6 mice that spanned genes known to flank *H3a*.[32] To scan the YACs for whether any of them contain *H3a*, we retrofitted the YACs with a mammalian selectable marker and introduced each YAC stably into antigen-negative cells by the process of spheroplast fusion. This method worked remarkably well, and we were able to identify a YAC that carried the *H3a* gene. However, YACs are large cloning vectors that can carry over 1000 kb of genomic mouse DNA, so we still had a long way to go to identify the *H3a* gene.

The main value of identifying a YAC carrying *H3a* was to home-in on the genetic segment represented by the YAC. However, the critical resource for the identification of the *H3a* gene was a panel of mutant lymphoblastoid cell lines selected in vitro by the technique of immunoselection for the loss of the H3a^a antigen.[27,31,32] Some of these cell lines carried micro-deletions with deletional breakpoints flanking *H3a*. These localized *H3a* down to less than 200 kb, in an interval containing a novel, large (approx. 100 kb) candidate gene. Other H3a^a antigen loss mutant cell lines carried mutations that resulted in a specific loss of expression of this candidate gene. Cloning of its 8.5 kb transcript and transfection experiments confirmed that the gene encoded the H3a^a antigen.[37] Comparison of the sequences of this gene from H3a^a and H3a^b mouse strains suggested one avidly binding nonamer sequence with a conventional D^b-binding motif (Asparagine in position 5 and a Leucine in position 9) that also had amino acid change between the two common H3a alleles (Table 1.2). Corresponding synthetic nonamer peptides for the H3a^a allele (ASPCNSTVL) but not the H3a^b allele (TSPRNSTVL) sensitized target cells for lysis in the picomolar range. These results proved that ASPCNSTVL was the H3a^a antigen.

CTLs generated by immunizing *H3* congenic strains in the a anti-b direction recognized target cells coated with the TSPRNSTVL synthetic peptide corresponding to the H3a^b allele of the same nonamer. Thus, CTLs are generated reciprocally depending on non-conservative changes (Alanine <-> Threonine in P1, Cystine <-> Arginine in P4) within the same nonamer peptide. The 1,888 amino acid protein carrying the H3a antigens has a human homologue, three zinc finger motifs and an SH2 domain.[37] Independently, Salvadore et al have identified a partial cDNA from this gene.[38] The expressed protein appears to participate in the insulin receptor-mediated intracellular signal cascade where it interacts with the GRB-2 and FYN proteins. We have named this gene zinc finger 106 (Zfp106) rather than *H3a* in realization that its probable evolutionary biological function is in signal transduction rather than anything related to its antigenic activity.

Table 1.2. Antigens of the greater *H3* complex

Antigen	Allele	(A.A. Change)	Directionality	Mechanism
B2M	$B2m^a$	($\underline{D}^{85}$)	Reciprocal	Alters antigen presentation
	$B2m^b$	($\underline{V}^{85}$)		
H3a	$H3a^a$	(ASP$\underline{C}$NSTVL)	Reciprocal	Alters Tcr contact sites
	$H3a^b$	(TSP$\underline{R}$NSTVL)		
H13	$H13^a$	(SSV$\underline{V}$GVWYL)	Reciprocal	Alters Tcr contact sites
	$H13^b$	(SSV$\underline{I}$GVWYL)		
H3b	$H3b^b$	(Unknown)	Unknown	Unknown

H13, a Type I H Gene Whose Antigens Push the Limits of Self-Non-Self Discrimination

In Chapter 5, Mendoza and collaborators describe their remarkably efficient functional cloning approach and its use to characterize the *H13* gene. The novel aspects of *H13* are: (1) that the nonamer D^b-binding peptide does not abide to the conventional D^b-binding motif; (2) the molecular basis underlying the H13^a and H13^b alleles is a very conservative amino acid change (Valine<->Isoleucine); (3) in keeping with the conservative nature of the amino acid change responsible for antigenicity, CTLs show extensive cross reactivity with synthetic peptides corresponding to their self-epitope, but only at peptide concentrations higher than those normally encountered in vivo; and (4) H13^a and H13^b are reciprocally antigenic.[39] The function of the *H13* gene is not known.

H3b, a Type II Gene

The characterization of type II minor *H* genes, such as *H3b*, has lagged behind type I genes. We have narrowed *H3b* to a very small (>0.2 centiMorgan) genetic interval, and are proceeding to identify genes within this interval.[27,30] It should be possible to identify *H3b* by positional cloning and the testing of suggested candidate genes.

A Simple Model for the Greater H3 Complex Based on T-Cell Collaboration

Armed with an improved molecular understanding of the *H3* complex, what can one say regarding the histocompatibility phenotype that originally served as the basis for selection of *H3*-congenic mouse strains? Figure 1.4 summarizes results examining the ability to generate CTLs and skin graft survival of intra-*H3* congenic mouse strains that separate B2M and H3a, from H3b, and from H13. Skin grafts are rejected slowly if at all when donor and recipient differed at only a type I disparity (H3a/B2M or *H13*) and not at all when donor and recipient differed at only the type II locus, *H3b*. A similar pattern was observed for CTL generation after immunization with spleen cells. Mice immunized with cells from mice that differ only by type I antigens do not generate CTLs unless they are deliberately provided with a helper stimulus from male specific HY antigens (by immunizing female mice with male cells). Mice that are immunized with cells from mice that differ only by type II antigens also fail to generate CTLs, while mice that are immunized with cells differing by Type I + II antigens efficiently generate CTLs directed against the type I antigens.[40,41]

The explanation for these data is based on the well-established concept of T-cell collaboration: the phenotype of tissue rejection is the result of obligate collaboration between CTLs responding to products of type I loci and helper cells responding to products

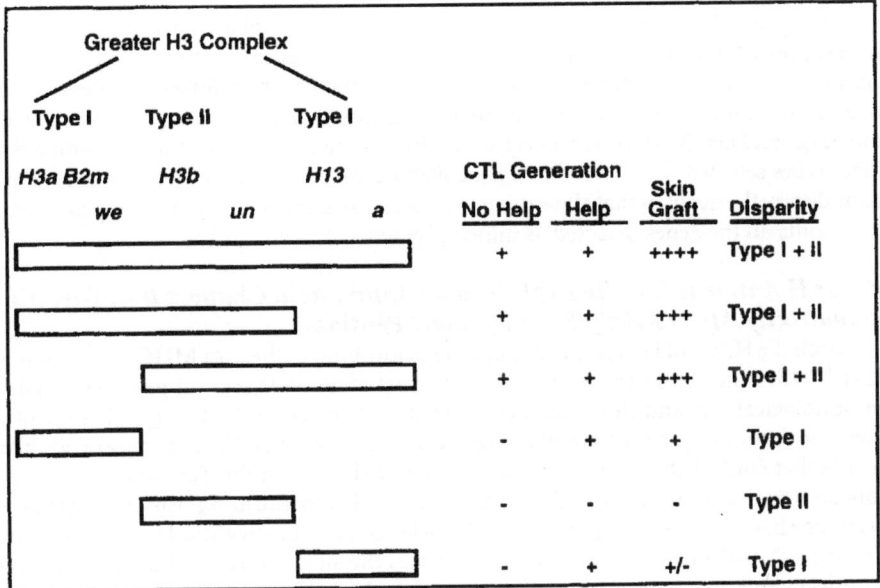

Fig. 1.4. Reinterpretation of the *H3* conundrum. "No Help" indicates that responder mice were immunized with sex-matched cells differing genetically as indicated. "Help" indicates that female mice were immunized with male cells differing genetically as indicated, thus providing a type II helper stimulus via HY antigens. + and – indicates the detection or absence of CTL by conventional cell mediated cytolysis assays. Intensity of skin graft rejection is indicated in scale of – (none) to ++++ (intense). Skin graft survival data from refs. 13,14,19,41. CML data and intra-*H3* recombinant strains from ref. 41.

of type II loci. CTLs induced without help reject allografts inefficiently because they are inefficient secretors of IL-2, while helper cells secrete copious IL-2.[42,43] Thus both a type I and type II disparity is required for full expression of the trait of graft rejection.[44]

It is important to note that donor skin differing at type II locus *H3b* alone failed to exhibit a rejection phenotype,[41] suggesting that helper T cells responding to *H3b* antigens are not efficient effectors of the allograft response. Instead, *H3b* expresses its immunological phenotype by enabling type I genes to be more readily detected. The fact that *H3b* maps between *H3a*, *B2M* and *H13* suggests that it may be the *H3*-linked *Ir* gene proposed many years earlier.[11-15]

Lessons Learned from *H3*

The intensive analysis to which the greater *H3* complex has been subjected provides a window into understanding cellular and molecular features of allogeneic rejection from the microcosm of minor H congenic strains, but the lessons learnt provide key insights into the larger issues of self-nonself discrimination, evolution, and clinical transplantation.

Among Many Genes, Few Are Chosen

On a genomic level, congenic segments are very large. The greater *H3* complex spans about 23 centiMorgans. About 70 distinct genes have been identified in this chromosomal segment. This number is likely a gross underestimate. Based on recombination frequency, which is a measure of gene linkage, the mouse chromosomes are comprised of about 1,600 centiMorgans. Assuming that there are about 100,000 genes in the mammalian genome, the greater *H3* complex could carry over 1,400 genes! In reality, not all of these genes are likely to be expressed in tumor or normal tissue grafts, but this estimate underscores the genetic complexity of transferred congenic chromosomal segments. The number of genes that could theoretically contribute to the trait of tissue rejection is therefore considerable. However, four *H* genes, *H3a, B2M, H3b,* and *H13,* are sufficient to account for the histocompatibility phenotypes selected for by Snell and his collaborators₃.[2,11,12] Therefore, few genes among many display the qualities that allow them to be detected as *H* genes. This imposes considerable constraints on the genes detected as minor *H* genes.

Minor H Antigens Can Be Explained by Amino Acid Changes that Directly or Indirectly Affect Avidly Binding Short Peptides

Both the H3a and H13 minor *H* peptides avidly bind to the host MHC class I molecule, H2-D^b. When they do so, the peptides become part of the wide array of peptides that define immunological self, and thus mice are tolerant to their self allelic form of these peptides. However, the bound peptides are alloantigenic to mice when they differ in amino acid changes in sites that contact the T-cell receptor (Tcr). The changes in the Tcr contact sites can be nonconservative, as are the position 1 Alanine <-> Threonine and position 4 Cystine <-> Arginine changes that distinguish a and b alleles of H3a, or very subtle, as is the position 4 Valine <-> Isoleucine change that distinguishes the alleles of H13. Mitochondrial minor H antigens and the human HA1 antigen can be explained by a similar mechanism,[47-50] (see Chapters 5 and 8).

Amino acid changes that indirectly influence the display of peptides bound to MHC proteins would also be detected as minor H peptides. β_2M is an novel example of such a mechanism. However, it underscores the fact that genes with amino acid changes that indirectly influence the display self-peptides could disguise as minor *H* loci. Additionally, at least one minor H antigen, H60 (previously referred to as CTT5), is determined by polymorphism at the transcriptional level.[51] Mice that fail to transcribe the *H60* gene respond to a K^b bound peptide derived from the precursor H60 protein.

Slow, Random and Nonsynomous Nucleotide Changes Give Rise to Minor H Peptides

Among common laboratory mouse strains, there appear to be only two alleles of the H3a and H13 nonamer peptides (unpublished observations).[39] (The former has a third allele in mice from a separate closely related species, *Mus spretus.*) Comparison of the sequences of *H3a* and *H13* genes with other genes from the same chromosome segment suggests that the degree of nucleotide variation is similar to autosomal genes that are not detected as minor *H* genes. These observations are in keeping with the previously predicted concept that minor H antigens arise from rare genetic changes in proteins with normal physiological functions arising within a species as a consequence of normal genetic drift.[52,53] The chance that these rare genetic changes will affect a peptide that avidly binds to the peptide cleft of a given allelic variant class I molecule is exponentially more rare. I believe that this simple exclusion principle is the primary rule that limits the number of minor *H* loci.

Given that amino acid changes leading to minor H peptides are infrequent and that each allelically variant MHC molecule has different peptide binding motifs, it is also clear that polymorphic peptides detected as a minor H antigen in the context of one MHC allele is unlikely to be detected as a minor H antigen in the context of another MHC allele. The

---

₃It is possible that different minor *H* loci inhabit the greater *H13* complex when congenic segments of varied donor strains are analyzed. For a more complicated interpretation of *H* loci in the greater *H3* complex, the reader is referred to studies of Graff and colleagues.[17,45,46]

genetic map shown in Figure 1.1 is based primarily on mice of the H2^b haplotype. Genetic maps of minor *H* loci in the context of different MHC haplotypes are likely to differ greatly.

Genes Encoding Class I and Class II Presented Minor H Antigens are Usually Distinct

The notion that type I and type II minor *H* genes are distinct was originally based on the analysis of *H3*, but now several additional examples strengthen this contention, including the *H46*, *H47* (formerly *H4*) gene complex on mouse chromosome 7,[54] and the genes encoding male-specific (HY) antigens[55] (see Chapter 6). There are two factors likely to contribute to this distinction. The first is based on the very limited amino acid variation of autosomal minor *H* genes. Given low sequence variation, the probability that a protein will coincidentally have amino acid changes in a peptide capable of binding to a specific allele of a class I molecule (thus acting as a type I antigen) and amino acid changes in a peptide that binds a specific class II MHC molecule (thus acting as a type II antigen) is low. The fact that class I-presented peptides theoretically could arise from any endogenously expressed gene while class II-presented peptides are generally limited to endosomally processed secreted or membrane proteins would tend to exaggerate the separation of type I and type II genes. Given these considerations, it is not surprising that type I and type II genes are usually distinct.

Many, but Not All, Minor H Congenic Strains Contain Both Type I and Type II Loci

Both historic and more recent data discussed in preceding sections support the concept that type I and type II minor *H* genes of the greater *H3* complex were coselected in congenic strains because both gene types were required for the rejection phenotype. Other minor H loci defined by congenic strains are likely to be similarly biased towards situations where both type I and type II genes are closely linked. These loci would give the appearance of being "strong" minor *H* loci and would readily generate CTL responses (e.g., *H1*, *H3*, *H4*, *H7*, and *H28*). The Y chromosome also contains type I and type II minor *H* genes confined to a short chromosomal segment that synergize to produce a strong rejection phenotype, but localization of these genes on the Y chromosome is the result of the unique evolution of this chromosome rather than experimental selection (see Scott et al, Chapter 6). Based on our experiences, type I minor *H* loci can elicit a very weak skin graft rejection phenotype; however, without a deliberate helper stimulus they are typically defective in eliciting CTLs following immunization with spleen cell. It is likely that professional antigen presenting cells, such as Langerhans cells, in some cases enable this weak rejection phenotype to be expressed in the absence of CD4$^+$ T-cell help. Some of the congenic strains produced by Bailey's very sensitive skin graft rejection technique (e.g., *H29* and *H34*) appear to be in the category of isolated type I loci.[41]

Can genetically isolated type II loci confer the trait of graft rejection? Although we have failed to observe this capacity for the *H3b* type II locus, studies of graft rejection induced by the HY antigen in the context of certain H2 haplotypes are consistent with CD4$^+$ cells causing rejection without CTLs.[56] An understanding of the molecular basis of these two antigens is the best way to resolve this issue.

Application of the Lessons Learned to Human Transplantation

Molecular analysis of the one autosomally encoded type I human minor *H* gene identified to date, *HA1*[50], agrees with the concept that autosomal minor *H* genes arise as a consequence of natural genetic variation of proteins whose natural function is not causally related to their rejection phenotype. As found in mice, one expects that type I and type II loci in humans will typically be encoded by separate genes. The range of genetic diversity among humans is not unlike that found among the wide array of mouse strains that are currently available. With the power of genetic manipulation and rapidly evolving knowledge of the molecular basis of mouse minor H antigens, the mouse model therefore offers an increasingly important surrogate to develop methods to ameliorate graft vs. host disease associated with heterologous HLA-matched in bone marrow transplants and chronic rejection associated solid tissue transplants.

Acknowledgments

I wish to express my deep appreciation to Robert Click, Allan Peter Davis, Greg Christianson, Aamir Zuberi, Larry Mobraaten, Nilabh Shastri, Lisa Mendoza, and Pedro Paz, all of whom contributed to these studies, and to the N.I.H. National Institute of Allergy and Infectious Diseases for providing me with the support to conduct the studies.

References

1. Snell GD. Methods for the study of histocompatibility genes. J Genetics 1948; 49:87-108.
2. Snell GD. Histocompatibility genes of the mouse. II. Production and analysis of isogenic resistant lines. J Natl Cancer Inst 1958; 21:843-877.
3. Snell GD. The genetics of tissue transplantation. In: Green EL, ed. Biology of the laboratory mouse. New York: McGraw Hill, 1966:457-491.
4. Gorer PA, Lyman S, Snell GD. Studies on the genetic and antigenic basis of tumor transplantation. Linkage between a histocompatibility gene and "fused" in mice. Proc Roy Soc B 1948; 135:499-505.
5. Counce S, Smith P, Barth R et al. Strong and weak histocompatibility differences in mice and their role in the rejection of homografts of tumors and skin. Ann Surg 1956; 144:198-204.
6. Bailey DW. Genetics of histocompatibility in mice I. New loci and congenic strains. Immunogenetics 1975; 2:249 -256.
7. Bailey DW. Recombinant inbred strains and bilineal congenic strains. In: Foster HL, Small JD, Fox JG, eds. The Mouse in Biomedical Research. Vol. 1. New York: Acad Press 198; 1:223-239.
8. Bailey DW, Mobraaten LE. Estimates of the number of loci contributing to the histoincompatibility between C57BL/6 and BALB/c strains of mice. Transplantation 1969; 7:394-400.
9. Graff RJ, Bailey DW. The non-H2 histocompatibility loci and their antigens. Transplant Rev 1973; 15:26-49.
10. Johnson L. How many histocompatibility loci do congenic mouse strains differ? J. Heredity 1981; 72:27-31.
11. Snell GD, Bunker HP. Histocompatibility genes of mice IV. The position of H3 in the fifth linkage group. Transplantation 1964; 2:743-751.
12. Snell GD, Cudkowicz G, Bunker HP. Histocompatibility genes of mice VII. H13, a new histocompatibility locus in the fifth linkage group. Transplantation 1967; 5:492-503.
13. Berriman JH, McKann CF. Transplantation immunity involving the H3 locus: Graft survival times. J Natl Cancer Inst 1960; 25:111-118.
14. Gasser DL. Genetic studies on the H3 region of mice and implications for polymorphism of histocompatibility loci. Immunogenetics 1976; 3:271-276.
15. Gasser DL. H3-linked unresponsiveness to Ea-1 and H13 antigens. In: Katz DH, Benacerraf B, eds. The Role of the Histocompatibility Complex in Immune Responses. New York: Acad Press, 1976:289-295.

16. Fierz W, Farmer GA, Sheena JH et al. Genetic analysis of the non-H2-linked *Ir* genes controlling the cytotoxic T-cell response to HY in H2^d mice. Immunogenetics 1982; 16:593-601.
17. Graff RJ, Hauptfeld V, Riordan K et al. Continued mapping of chromosome 2 genes. Immunogenetics 1994; 40:21-6.
18. Sadegh-Nasseri S, Kipp DE, Taylor BA et al. Selective reversal of H2 linked genetic unresponsiveness to lysozymes I. Non-*H2* gene(s) closely linked to the Ir-2 locus on chromosome 2 permit(s) an antilysozyme response in H2^b mice. Immungenetics 1984; 20:535-546.
19. Graff RJ, Brown D, Snell GD. The alleles of the H13 locus. Immunogenetics 1978; 7:413-423.
20. Michaelson J. Genetics of beta-2 microglobulin in the mouse. Immunogenetics 1983; 17:219-259.
21. Serreze D V, Bridgett M, Chapman HD et al. J Immunol 1998; 160:1472-1478
22. Bevan MJ. The major histocompatibility complex determines susceptibility to cytotoxic T cells directed against minor histocompatibility antigens. J Exp Med 1975; 142:1349-1364.
23. Gordon RD, Simpson E, Samelson LE. In vitro cell-mediated immune response to the male-specific (HY) antigen. J Exp Med 1975; 142:1108-1120.
24. Steinmuller D. Which T cells cause allograft rejection? Transplantation 1985; 40:299-233.
25. Roopenian DC, Click RE. A new cytotoxic lymphocyte-defined antigen coded for by a gene closely linked to the H-3 locus. Immunogenetics 1980; 10:333-34.
26. Click RE, Schneider D, Roopenian DC. A new minor histocompatibility locus linked to H3. J Immunol 1981; 126:2378-2381.
27. Zuberi AR, Nguyen HQ, Auman HJ et al. A high resolution genetic linkage map of mouse chromosome 2 extending from thrombospondin to paired box gene 1 including the H3 mouse minor histocompatibility complex. Genomics 1996; 33:75-84.
28. Roopenian DC, Anderson PS. Generation of helper cell-independent cytotoxic T lymphocytes is dependent upon L3T4$^+$ helper cells. J Immunol 1988; 141:391-7.
29. Roopenian DC, Anderson PS. Adoptive immunity in immune deficient scid/scid mice. I. Differential requirements of naive and primed lymphocytes for CD4$^+$ T cells during rejection of minor histocompatibility antigen disparate skin grafts. Transplantation 1988; 46:899-904.
30. Zuberi AR, Roopenian DC. High resolution mapping of a minor histocompatibility antigen gene on mouse chromosome 2. Mammal Genome 1993; 4:516-522.
31. Zuberi AR, Dudley ME, Christianson GJ, Roopenian DC. Gene mapping in a murine cell line by immunoseletion with cytotoxic T lymphocytes. Genomics 1994; 19:273-279.
32. Zuberi AR, Christianson GJ, Dave SB et al. Expression screening of overlapping yeast artificial chromosomes identifies a clone that carries the mouse H3a minor histocompatibility antigen gene. J Immunol 1998; 161:821-828.
33. Rammensee HG, Robinson PJ, Chrisanti A et al. Restricted recognition of β$_2$ microglobulin by cytotoxic T lymphocytes. Nature 1986; 319:502-4.
34. Kurtz ME, Martin-Morgan D, Graff RJ. Recognition of the β$_2$ microglobulin-B molecule by a CTL clone. J Immunol 1987; 138:87-90.
35. Pérarnau B, Siegrist CA, Gillet A et al. β$_2$-microglobulin restriction of antigen presentation. Nature 1990; 346:751-4.
36. Hederer RA, Chandler PR, Dyson PJ et al. Acceptance of skin grafts between mice bearing different allelic forms of beta 2-microglobulin. Transplantation 1996; 61:299-304.
37. Zuberi AR, Christianson G J, Mendoza LM et al. Positional cloning and molecular characterization of an immunodominant cytotoxic determinant of the mouse H3 minor histocompatibility complex. Immunity 1998; 9:687-698.
38. Salvatore P, Hanash CR, Kido Y et al. Identification of sirm, a novel insulin-regulated SH3 binding protein that associates with Grb-2 and FYN. J Biol Chem 1998; 273:6989-6997.
39. Mendoza LM, Paz P, Zuberi AR et al. Minors held by majors. The H13 minor histocompatibility locus defined as a peptide/MHC class I complex. Immunity 1997; 7:461-472.

40. Roopenian DC, Davis AP. Responses against antigens encoded by the H3 histocompatibility locus: Antigens stimulating class I MHC and class II MHC restricted T cells are encoded by separate genes. Immunogenetics 1989; 30:335-43.

41. Roopenian DC, Davis AP, Christianson GC et al. The functional basis of minor histocompatibility loci. J Immunol 1993; 1561:4595-4650.

42. Roopenian DC, Orosz CG, Bach FH. Responses against single minor histocompatibility antigens. II. Analysis of cloned helper T cells. J Immunol 1984; 132:1080-1084.

43. Roopenian DC, Widmer MB, Orosz CG et al. Response against single minor histocompatibility antigens. I. Functional and immunogenetic analysis of cloned cytolytic T cells. J Immunol 1983; 131:2135-2140.

44. Roopenian DC. What are minor histocompatibility loci? A new look at an old question. Immunol Today 1992; 13:7-10.

45. Graff RJ, Martin-Morgan D, Kurtz ME. Multiplicity of chromosome 2 histocompatibility genes: New loci, H44 and H45. Immunogenetics 1987; 26:111-114.

46. Graff RJ, Kurtz ME, Paul R, Martin D et al. Additional mapping of mouse chromosome 2 genes. Immunogenetics 1991; 33:96-199.

47. Loveland BE, Wang CR, Yonekawa H et al. Maternally transmitted histocompatibility antigen of mice: A hydrophobic peptide of a mitochondrially encoded protein. Cell 1990; 60:971-80.

48. Morse M C, Bleau G, Dabhi VM et al. The COI mitochondrial gene encodes a minor histocompatibility antigen presented by H2-M3. Immunol 1996; 156:3301-7.

49. Bhuyan PK, Young LL, Lindahl KF et al. Identification of the rat maternally transmitted minor histocompatibility antigen. J Immunol 1997; 158:3753-60.

50. den Haan JM, Meadows LM, Wang W et al. The minor histocompatibility antigen HA1: A diallelic gene with a single amino acid polymorphism. Science 1998; 279:1054-1057.

51. Malarkannan S, Shih P, Eden P et al. The molecular and functional characterization of a dominant minor H antigen, H60. J Immunol 1998; 161: 3501-3509.

52. Townsend ARM, Rothbard J, Gotch FM et al. The epitopes of influenza nucleoprotein recognized by cytotoxic T lymphocytes can be defined with short synthetic peptides. Cell 1986; 44:959-968.

53. Roopenian DC, Christianson GJ, Davis AP et al. The genetic origin of minor histocompatibility antigens. Immunogenetics 1993; 38:131-140.

54. Davis AP, Roopenian DC. Complexity at the mouse minor histocompatibility locus H4. Immunogenetics 1990; 31:7-12.

55. King TR, Christianson GJ, Mitchell MM et al. Deletion mapping by immunoselection against the H-Y histocompatibility antigen further resolves the Sxr-a region of the mouse Y chromosome and reveals complexity of the Hya locus. Genomics 1994; 24:159-68.

56. Hurme M, Chandler PR, Hetherington CM et al. Cytotoxic T-cell responses to HY: Correlation with the rejection of syngeneic male skin grafts. J Exp Med 1978; 147: 768-775.

CHAPTER 2

Sizing Up the Set of *H* Genes in Mice

Donald W. Bailey

Early studies in histocompatibility focused on the number of genes that evoke graft rejection. Initially, (*H*) genes were treated together as a set and the size of this set was the most important question. Not until George Snell isolated *H* genes in congenic strains[1] could an individual *H* gene's function be studied—as amply demonstrated with the *H2* complex. Excluding *H2* for now, subsequent studies on histocompatibility genetics still have an impact on our concept of the size of the *H* gene set even though those studies might have been designed to reveal other properties. In this limited review I want to concentrate on *H* set size; I find that there are many good reasons for H-set size to be more extensive than early studies indicated. [For more comprehensive reviews of minor *H* genes see refs. 2,3.] I also want to speculate on how *H* genes fit into the scheme of things; just what should we expect *H* gene functions to be? To give perspective as well as a personal touch, I will relate how the results from various *H* gene studies as they were encountered caused a research project in which I was involved to be modified as that project unfolded during the 1960s and 1970s.

The Basis of Transplantation

The research project, which I shall describe below, was highly dependent on an important discovery made early this century by Little and Tyzzer.[4] Their experiment was simply to transplant tumor tissue originating in the inbred Japanese waltzing mouse strain to the F2 generation produced by crossing that strain to the dba (now DBA) inbred strain. These strains had been unintentionally inbred in order to maintain their recessive genotypes. Little and Tyzzer had predicted that a small percentage of the F2 mice would die from transplanted tumor tissue, a prediction based on a genetic hypothesis that Little had spelled out earlier.[5] They reasoned that a large set of codominant genes, expressed in the tumor, was responsible and that the greater the number of genes interfering with the growth of the transplant, the fewer recipient mice would die from the transplant. Little and Tyzzer made it very clear how their hypothesis differed from others, especially that developed by Loeb,[6] which was based on the genetic relationship of donor and host but not at the gene level. Derived from the Little-Tyzzer hypothesis was an equation by which they estimated the number of effective genes to be 12 to 14. *H* gene set sizing began here.

Importantly, the genetic model could be generalized. A few years later Little and Johnson showed that this concept applied to normal tissue (splenic) grafts as well.[7] Most confirmatory studies of this genetic model during the 1920s and 1930s, however, concentrated on tumor grafts applied to yet other inbred strains.[8,9]

Minor Histocompatibility Antigens: From the Laboratory to the Clinic, edited by Derry Roopenian and Elizabeth Simpson. ©2000 Landes Bioscience.

The legacy of Little and Tyzzer's work has been threefold:
1. The genetic basis of transplantation was established and has been upheld ever since.
2. The importance of inbred strains for studies of transplantation and immunology in general was established.
3. The research theme on which the Jackson Laboratory was founded by Little and on which the Laboratory has subsequently thrived was originated.

Leonell Strong, another pioneer in tumor transplantation studies, was also a promoter of inbred mouse strains and was instrumental in establishing them at the Jackson Laboratory. According to him not all geneticists were sanguine about the inbreeding of mice.[9] [Ironically, William Castle, considered the father of mammalian genetics, was in this latter group.] First, those of this mind thought it would not be successful because of demonstrated debilitating effects of inbreeding in other animals. Also, a few of this group thought that incestuous matings forced on mice were immoral.

Snell's Laws

What Little and Tyzzer found led directly to Snell's Laws of transplantation. These laws express what one would expect from codominant immunogen-producing genes that behave in a Mendelian fashion. Snell listed five laws;[10] for brevity I have condensed them to three:
1. Tissue grafts exchanged between genetically identical mice, i.e.,within inbred strains or within an interstrain F1, will survive.
2. Grafts from any mouse in a generation derived solely from crossing two fully inbred strains, no matter from what later generation, will survive on the F1 host, the universal recipient.
3. Grafts exchanged otherwise will likely be rejected, with the probability of rejection depending upon the mismatching of *H* genes in donor and/or host.

In short time Snell's Laws had to be modified. Eichwald and Silmser, using skin grafts, were surprised to find females rejected male grafts within inbred strains.[11] This was quickly recognized by Hauschka[12] and Snell[13] likely to be caused by a gene on the Y chromosome. Thus, Snell's Laws were in need of modification but were still consistent with Mendelian principles, which meant that intra-strain grafts thereupon were expected to succeed only if grafts were exchanged between like-sexed mice.

Grafting Methodology

Tumor grafting gradually gave way to skin grafting as the preferred assay procedure for studies in transplantation genetics. The skin-grafting technique was developed and applied to various laboratory species by Billingham and Medawar.[14] This freed the immunogeneticist from reliance on tumor grafts fettered with their inherent problems of chromosome loss, proneness to infections, and lowered sensitivity to the rejection process. The new technique involved placing a skin graft onto the flank of the recipient mouse and protecting it by a plaster cast. Later on a more expedient technique was developed to place tailskin orthotopically and protect it by glass tubing for continuous viewability.[15] Skin grafting, and especially the grafting of smaller tailskin slices, increased the sensitivity of the test system. Thereby, weaker histoincompatibilities have been revealed and larger estimates of *H*-gene number obtained.

H-Mutation Project

I was able to convince Henry Kohn, a radiation biologist at the University of California Medical Center in San Francisco, of the considerable potential in the study of both spontaneous and induced *H* gene mutations in mice. Thereupon, he invited me—along with several of my mouse strains under development—to join his group. At that time, 1961, there was a paucity of data on induced mutation rates in mammals; there was a need to extend the variety of genes examined in order to have a more representative sampling of the mouse genome. *H* genes promised to help fulfill that need.

We were convinced that the set of *H* genes would be useful for assaying the mutagenic effects of irradiation because there was:

1. a potentially large number of genes in that set (the number was thought to be about 14 at that time),
2. a good possibility of identifying the locus of a mutated gene by the specificity of the gene product, and by its location on the linkage map, and
3. a likelihood of there being but few viability effects to vitiate the results, as indicated by the fact that these genes were highly polymorphic. [In the early 1960s, only a few types of genes were recognized as being polymorphic.[16]]

To attain our research objective, we planned a four-pronged attack:

1. to assay a large population of isogenic mice for *H* gene mutations.
2. to create a pertinent set of congenic strains for identifying each mutated locus,
3. to make more accurate estimates of the number of *H* genes with which we would be working, and
4. to devise a quick way of finding the genetic linkages of mutant *H* genes as they arose. We continued in this quest despite the failure of Godfrey and Searle in their pioneering effort to find *H* gene mutations in mice.[17]

We applied the modified Snell's Laws in our project. We simply exchanged tailskin grafts amongst like-sexed, reciprocal-type F1 hybrids of the inbred strains BALB/cBy [C] and C57BL/6By [B6]. Such a graft should be rejected only if it carried an immunologically detectable H mutation.

The paternal mice in these crosses were either irradiated with gamma rays, or not. Orthotopic tailskin grafts[15,18] were exchanged between mice in a strategic pattern: each F1 mouse served as both recipient and donor to two other mice, thereby permitting the classification of each mutation as a gain, a loss or both a gain and a loss of immunogenicity.[19]

Genetically-Defined Strains

Congenic Strains

The system within which the whole project was to be confined was that defined by the C and B6 strain genomes. Following Snell's Laws, it would be those *H* loci at which these two strains differ that would allow us to identify and locate mutant loci in the murine genome. Therefore, we began to develop a set of congenic strains that represented these *H* loci.[20]

At that time Snell had developed or was developing an array of congenic lines, which he at first called isogenic resistant (IR) lines and later, congenic resistant (CR) lines.[1] Why didn't we use those ready-made lines? Unfortunately for us, Snell's objective had been to maximize the range of *H* genes to be 'captured.' He had introduced *H* genes from miscellaneous stocks onto several inbred backgrounds; we needed bilineal congenic strains, i.e., strains derived from only two strains, both highly inbred.

Snell's strains differed in yet other ways: he used tumor tissue as test grafts because he felt such grafts were technically more easily performed than skin grafts.[13] Snell selected for the absence of an *H* gene product in the recipient mouse and not for the presence of a product in the donor mouse, i.e., he selected for the host's "resistance" to the tumor. This forced him to double the number of generations required to attain the desired genetic background purity because—following Snell's Laws—there had to be alternating test generations for selecting homozygous parents that rejected the tumor graft.

Snell wrote about how he became interested in developing congenic strains.[21] Seeing this account and observing his work firsthand, I sense his greatness to be grounded in his persistent pursuit of the single objective, the dissection of the set of *H* genes through the manipulative procedure of generating congenic strains. Interestingly, Snell found Little not to be enthusiastic about this project.

Recombinant Inbred Strains

We decided that a means of distinguishing the mutated *H* gene would be by its location on the linkage map, and a quick way to locate a gene would be to employ a set of recombinant inbred (RI) strains, the CXB set, which I had begun at NIH Bethesda, MD, in 1959 just prior to beginning this mutation study.[22,23] Since this RI set was derived from the B6 and C strains, it would fit in well with our scheme of analysis. The plan was first by use of our RI strains to map the *H* genes that had been isolated in our congenic lines, then by allelism tests against those same isolated *H* genes to identify each mutated *H* locus.

Mutation Study Outcome

"The best laid schemes o' mice an' men Gang aft agley." -R. Burns

The HX Gene

In our search for mutations, it wasn't long before we started seeing many 'mutants.' Upon closer study these were observed whenever the grafts were exchanged between males of reciprocal F1 hybrids. These soon became clearly explicable as due to one or more *H* genes on the X chromosome.[24] Upon this discovery, F1 mice were no longer universal recipients; grafts would not survive when exchanged between reciprocal-type F1-hybrid males. Snell's Laws had to be modified again.

However, what about female grafts on males? We hadn't made such grafts in the routine assays for mutants. A few years earlier, Mary Lyon proposed a hypothesis which asserted the inactivation of either the paternal or the maternal X chromosome randomly in each somatic cell in female mammals.[25] We could now test this hypothesis with *HX*. And indeed, F1-female skin was rejected in a mosaic fashion when placed on F1-male hosts.[18] This was evidence that the *HX* gene(s) had its effect at the cellular level.

In one of his studies on tumor biology, Strong felt he had chanced upon evidence of a sex-linked gene affecting tumor survival.[26] This would have been the first encounter with HX. However, it is not at all clear in that paper just what led him to that conclusion.

Considering these amendments to Snell's Laws, I find that while still keeping Mendelian genetics in mind, the Laws can be stated more simply:
"Havenots will react against the Haves"—just as they do in human society.

Types of H Mutation

After several replications of the study, we found that the vast majority of mutants were of the gain type which indicated they occurred at *H* loci other than those at which the two strains differ.[27] Thus, our plan to identify *H* mutants by exploiting newly constructed congenic strains failed, except for a few mutants, and thereby our plan to locate them on the linkage map failed. [Out of 125 mutants, 121 were gain, 3 were gain and loss, and 1 was a loss type.]

Irradiation Effects

Furthermore, we found that gamma irradiation of the fathers did not increase the mutation rate as it had for other types of genes. Later on, after Henry Kohn transferred his work to Harvard Medical School in Boston, he and Melvold also found x-irradiation failed to mutate *H* genes.[28] Our assumptions about finding mutations were grossly wrong. This was interpreted as the mutant event possibly being due to viral-genome incorporation.[27] However, looking back now, a more likely interpretation would be that gamma irradiation was too harsh and that any irradiation effect probably resulted in lethality, that most *H* genes—whatever their functions—are essential for viability of the mouse. This interpretation is supported by the finding by Kohn that application of triethylene melamine, a mutagen that produces point mutations, did increase the rate of *H* gene mutation.[29]

Replicate Studies

Finally, the mutation rate differed significantly between replicate experiments. In five replicated experiments performed in San Francisco, the spontaneous mutation rates for the *H* set varied from 0.006-0.072 per gamete, a statistically significant difference.[27] Moreover, in experiments replicated later under the same protocol by Kohn and Melvold in Boston, the spontaneous mutation rate was 0.002 per gamete.[30] This again was a statistically significant difference from those obtained in experiments performed in San Francisco.

These differences were indeed troublesome. How can we account for them? A portion were obviously due to our failure in the early experiments to pre-screen mice to be used as parents and to exclude those found to be carriers of *H* gene mutations that had originated in earlier generations. [The size of this mutant gene pool with a continuous turnover in highly inbred strains—and thus always a potential threat to research plans—is predictable.][31] After adopting the prescreening procedure, our rate stayed near 0.010 per gamete.

Another portion of this variation in spontaneous mutation rate must be attributed to the nature of the testing procedure; it depends upon the sensitivity of the host mouse in its detecting and responding to the immunogen. As suggested by Kohn and Melvold, the immune response is expected to be quite sensitive to environmental factors such as nutrition, noise and handling by caretakers under which the mice are maintained.[30] Relevantly perhaps, we noticed a reduction in the intensity of the immune response to the HX difference upon our move from San Francisco to Bar Harbor (unpublished). The basis of this observed change remains unexplored.

Snell's Laws Challenged

Many experiments have been based on Snell's Laws. Our search for mutations was especially dependent on them. So when Hildemann and Cooper[32] published results of an experiment that refuted these laws we were devastated. Using A/J and B6 strains those authors found rejections of skin grafts from F2 and F3-generation donors on F1-generation hosts, a flagrant violation of the Laws. In fact they reported 53% of the F3-generation grafts to be rejected. If upheld, this would have wrecked our whole endeavor, but more importantly, Little and Tyzzer's genetic hypothesis would have been falsified.

We immediately repeated Hildemann and Cooper's experiment but using our strains (B6 and C) and could not corroborate their results.[33] Incompatibilities did not occur at a frequency above what we had observed in our mutation experiments. Moreover, we also felt that we surely would have observed such an increase in such events during our breeding regimens for developing bilineal congenic strains as well as our RI strains, and we did not. Several explanations were offered for the effects they observed, but after hearing by private communication that they had mouse colony health problems during the course of that experiment, we concluded that graft rejection was most likely caused by infected skin.

Estimating H Set Size

The size of the set of *H* genes can be estimated from different types of data.[34] They all tend to give underestimates for various reasons, some already mentioned. Most approaches have been improved upon by one means or another, but we shall see that the more we discover about the properties of *H* gene products, the more we realize we are underestimating *H* set size.

Segregational Data

After Little and Tyzzer, a number of studies were performed in different laboratories through the 1930s and 1940s using tumor grafting techniques to estimate *H*-gene number. Snell reviewed the published estimates of that time, and concluded that, in comparing any two inbred strains, the average number responsible for tumor graft rejection was close to seven.[1] He considered each locus to have a dominant allele (H) and a recessive allele (h), and therefore the genes that caused a rejection of a tumor from one parental strain were a different set than those that caused a rejection from the other parental strain. Therefore he doubled the number of *H* loci to 14. He did add, as an aside, that there might well be multiple alleles at these loci.

Genetic Dilution

However, these studies had a basic problem: their methodology was insensitive to making accurate estimates. There was such a large number of loci that few grafts would survive thereby causing the statistical error of estimate to be great. Little and Tyzzer realized this problem in their studies. Gilmour emphasized this point in reviewing various published estimates of *H* gene number.[35]

Due to this, we took a new tack by deliberately reducing the number of *H*-gene differences in the test mice. The backcrossing procedure used in producing bilineal-congenic strains was equivalent to a series of doubling dilutions. This would not only reduce the number of *H* genes in the donor mice, it would also reduce to a small degree the effects of linkage. We grafted tailskin from each of the incipient congenic strains onto the B6 and pooled the estimates from all incipient congenic strains to give a reliable estimate of the B6-C difference of 29 *H* loci.[36] We also exchanged grafts amongst mice within each of the incipient RI strains of the CXB set. After picking the appropriate generation and pooling the estimates of all RI strains, we estimated the B6-C difference to be 28.[36]

Graff and Brown made estimates of *H* gene differences between other strains using similar H set dilution and grafting techniques.[37] Strain differences were all in the same range as above with the highest estimate of 34. When the ploy of preimmunization was applied, the estimate more than doubled—from 27 to 63 for the B10.D2-DBA/2 strain difference.

Isolation

The ultimate application of segregational analysis is the isolation of *H* genes individually in congenic strains. The number of named *H* genes is now at 59[38] to which can be added the mitochondrial *H* genes.[3] Of course we cannot be assured each isolation involves a single gene; there may be a block of linked genes left in the introduced chromosomal segment. This has been demonstrated a number of times: the best known is that of the *H2* complex, and there are other examples as well.[3,39] So once again our estimate is too low.

Mutational Data

Mutations offer a means to estimate gene number independently of genetic segregation. Estimates so derived are not subject to underestimation due to linkage effects but are still dependent upon the sensitivity of the grafting regimen, as discussed above.

The overwhelmingly greater portion of gain-type over loss-type mutations indicated to us that most of these events occurred at loci outside of those at which the two strains differ. We concluded from this observation that the total number of *H* loci within the species is much greater in number than that number at which any two strains differ. After making some simplifying assumptions, we estimated that the number of *H* loci in *Mus musculus* would be at least 430.[34]

We can look at the mutation data on yet another basis.[34] Assume that *H* genes mutate at the same rate commonly cited for other genes in mammals (0.00001/locus/gamete). Now, if we take the mutation rate for the entire set of *H* genes to be 0.01/gamete—as we had found at the Medical Center in San Francisco—then the number of *H* loci in that block would be 1000. If we take the rate to be 0.0026 as found by Kohn and Melvold at Harvard,[30] the number would be 260.

The estimates based on inter-strain comparisons and those based on mutational data seem quite disparate, but in the final section of this chapter I will show how they may very well be reconcilable.

Hidden *H* Genes

Cumulative Effects

Graff et al showed that *H* genes have cumulative effects, i.e., the more *H* gene differences possessed by the donor and not by the host, the shorter the graft rejection time.[40] Moreover, it is noteworthy that the combined differences of a conglomerate of non-H2 genes without the presence of an H2 difference, evoke an immune response every bit as strong as a difference determined by H2 alone.[41] This suggests that *H* gene differences when few in number sometimes may remain undetected and unnamed. In a genetically segregating population, however, *H* genes that are immunologically weak when alone or few in number may by chance show up in numbers sufficient to effect graft rejection that was unpredicted by tissue typing of immunologically strong *H* genes. This cumulative effect, of course, would cause the estimates based on segregation or mutation to be too low.

This effect would also tend to influence the composition of congenic strains. The parents selected each succeeding backcross generation would tend to be those carrying differences of several *H* genes and especially those closely linked to form 'nests'.

Immunodominance

Immunodominance is the phenomenon in which one immunogen stimulates the immune response with concomitant loss of response to other immunogens expressed in the same cell.[42] This effect is the opposite of the cumulative effect. Although the effect results in hidden immunogens, I can see no way that it has affected estimates of *H* gene number. Other things being equal, an *H* gene whose product is either dominating or dominated in a segregating population would be included in counts derived from equations used for estimation, and a mutation to either type of gene would have been detected in our mutation study.

Gene Interaction

In our H mutation studies we observed that skin grafts from the B6 strain male were not rejected by the CB6F1 male host as soon as were those grafts from the B6CF1 male; both types of donor males with HXs from the same source but HYs and half of their autosomes from different sources.[43] This effect seems to be related to the effects observed subsequently by others where a gene closely linked to H2 modifies HY immunogenicity.[44-46] This type of dependency on other genes for expression, if widespread, would lead to an underestimated

H gene number, and also would likely affect the types of genes selected in developing congenic lines. In this regard, Roopenian (chapter 1) has suggested that the strongest skin graft rejection is caused by linked H genes whose products stimulate different but interdependent arms of the immune system.[47] According to his model, H genes are generally separated into one type whose product stimulates a helper T-cell (TH) response and another whose product stimulates a cytotoxic T-lymphocyte (CTL) response (reviewed in ref. 39). The presence of both H gene types greatly facilitates graft rejection.

If we accept this paradigm, each mutant incompatibility event—accepting these as true genetic mutational events—in our studies would have necessitated two mutations to have occurred in the same gamete. The rate of mutation observed in our San Francisco study (0.01), then, would be the square of the actual rate, and therefore the actual rate would be 0.1 per gamete. This, in turn, would mean either H loci are especially mutable or the number of H loci is much larger than suggested before and our estimates are by far too low.

However, some of our earlier efforts bear on the validity of this conclusion: If each mutant mouse carried two (usually unlinked) mutant genes, when we were producing congenic mutant H gene lines we would have noticed the lowered proportion of incompatible mice each succeeding backcross generation.[19] A third of our mutants behaved in this way, a result that we have attributed to their having weaker immunogenicity.

Roopenian suggests as a possible reconciliation that H mutants would more likely be detected in an environment in which the mice were immunologically challenged with viruses and/or microbes.[39] Environmental antigens are hypothesized to stimulate TH cells to help CTL cells in responding to the H mutant graft. As mentioned previously, we noticed a reduction in the intensity of the immune response to the HX difference upon our move from San Francisco to Bar Harbor. Environmental differences could explain this change. However, if environment played a key role in the detection of H mutations, grafts from congenic H mutant lines might loose their immunogenicity upon being treated with antibiotics or upon those lines being brought into a cleaner environment. We have not seen either of these outcomes (unpublished). The role of infectious organisms in the detection of H mutations thus remains to be established.

H2 Presentation Differences

The strength of graft rejection has been shown to be affected not only by genes that determine the immunogen(s) of the donor but also by genes that determine the host response. Especially noteworthy are the responses determined by the H2 complex toward HY and other non-H2 H gene products, as well as by other immune response genes and Chapter 6.[2,3]

If each haplotype of the MHC is capable of presenting its own repertoire of peptides, mice with each H2 haplotype can be expected to respond to a different, although possibly overlapping, subset of H-gene products. Our mutation experiments were conducted in F1 mice with both $H2^b$ and $H2^d$ haplotypes,[19] so it is likely that more mutations were detected than would have been in found in the context of a single H2 haplotype.

The H Gene Redefined

H gene product immunogenicity, per se, probably has no natural functionality. To me, an H gene is a gene whose product evokes an immunological rejection of a graft in vivo. With the development of new laboratory techniques, the in vitro CTL response is accepted as defining an H gene.[3] But Roopenian has shown that a CTL response, by itself, is not always adequate for a full rejection; it also needs to be abetted by a T helper cell.[39] Thus, the criterion for recognizing an H gene is evolving and an even greater number of loci will likely be eligible for membership. That said, the above estimates would again be too low.

Tissue Distribution

My above definition of the *H* gene does not specify the type of grafted tissue in which the immunogen is expressed. Each tissue can be expected to have its own subset of expressed *H* genes, which overlaps the subsets of other tissues.[48,49] This can only add to the *H* set size.

Significance of Set Size

What does this all mean? Does it matter that there is such a large *H* set? Perhaps yes, perhaps no. The *H* genes that matter are those that are polymorphic. They are the genes that affect transplant survival. All the others—excepting for mutation—are 'silent'.[50] Therefore estimates of interstrain differences are more pertinent to our interests than are estimates from mutations because they are more closely related to the *H* polymorphism of the species.

How well does an inbred strain represent the species? In an idealized case, an independently derived inbred strain from a randomly mating population can be considered, although diploid, genotypically equivalent to a gamete randomly drawn from that population. The *H* loci at which two such inbred strains differ, then, would be equivalent to those *H* loci that would be heterozygous in an individual from that same population. The problem with this model is that, first, *Mus musculus* is not a randomly mating population; it has a deme and subspecific structure. And second, the inbred strains were not independently derived; they came largely from mouse fancier's colonies, the progenitors of which were a small sample and not randomly drawn from the species.[51] Laboratory mice apparently originated by fanciers from the two subspecies *domesticus* (maternally) and *musculus* (paternally).

However, for a first approximation let us assume the inter-strain difference to be an estimate of the number of heterozygous *H* loci in an individual. This will not only be the simplest but also the most conservative assumption. From this we can estimate the limits that the number (L) of polymorphic *H* loci might have. The number depends on the allele frequencies at each of the polymorphic *H* loci. Let us take the simplest, most conservative case where there are only two alleles at each polymorphic *H* locus, and allele frequency (p) is the same for all such loci. Also, let us take the inter-strain difference conservatively to be 30. Accordingly, if p is 0.5, L is 60; if p = 0.1, L = 150; and if p = 0.01, L = 1500. [L is found by dividing 30 by 2p(1-p).] By this reasoning, the number of H loci in the species is at least twice the estimates coming from inter-strain differences. The real set of polymorphic H loci would no doubt show a spectrum of frequencies and each locus probably has more than two alleles. The published studies[52,53] on number of H alleles at a locus are too limited in scope to be useful here, for they are narrowly based on polymorphic loci identified in inbred strains. I suspect the actual frequencies in the species would tend to be at the extremes of the range rather than near 0.5. Thus, the number of *H* loci—even when we consider only the polymorphic ones—once again seems to approach the number estimated from the mutational data.

I see no reason to suspect the genome of *Homo sapiens* to be any less variable than that of *Mus musculus*. The breeding structure of our species is far from random; ethnic groupings of man would tend to maintain augmented genetic variation in a manner similar to the deme structure of mice. So, let us not be surprised to find our species to be polymorphic at perhaps hundreds of *H* loci. This presages for the transplant clinic how overwhelming the task might be in identifying *H* genes to be avoided in specific donor-host combinations.

References

1. Snell GD. Methods for the study of histocompatibility genes J Genet 1948; 49:87-108.
2. Loveland BE, Simpson E. The non-MHC transplantation antigens: Neither weak nor minor. Immunology Today 1986;7:223-229.

3. Loveland BE, Lindahl KF. The definition and expression of minor histocompatibility antigens. In: McCluskey J, ed. Antigen processing and Recognition. Boca Raton: CRC Press, 1991:173-192.

4. Little CC, Tyzzer EE. Further studies on inheritance of susceptibility to a transplantable tumor of Japanese waltzing mice. J Med Res 1916; 33:393-425.

5. Little CC. A possible Mendelian explanation for a type of inheritance apparently non-Mendelian in nature. Science 1914; 40:904-906.

6. Loeb L. The Biological Basis of Individuality. Baltimore: CC Thomas, 1945.

7. Little CC, Johnson BW. The inheritance of susceptibility to implants of splenic tissue in mice. Proc Soc Exp Biol Med 1922; 19:163-167.

8. Little CC. Genetics and the cancer problem. In: Dunn LC, ed. Genetics in the Twentieth Century. New York: Macmillan, 1951:431-472.

9. Strong LC. Inbred mice in science. In: Morse HC III, ed. Origins of Inbred Mice. New York: Acad Press, 1978:45-66.

10. Snell, JD, Stimpfling, JH. Genetics of tissue transplantation. In: Green EL, ed. Biology of the Laboratory Mouse. 2nd ed. New York: McGraw Hill, 1966:457-491.

11. Eichwald EJ, Silmser CR. (Note without title.) Transpl Bull 1955; 2:148-149.

12. Hauschka TS. Probable Y-linkage of a histocompatability gene. Transpl Bull 1955; 2:154-155.

13. Snell GD. A comment on Eichwald and Silmser's communication. Transpl Bull 1955; 3:29-31.

14. Billingham RE, Medawar PB. The technique of free skin grafting in mammals. J Exp Biol 1951; 28:385-340.

15. Bailey DW, Usama B. A rapid method of grafting skin on tails of mice. Transpl Bull 1960; 7:424-425.

16. Hutton JJ. Biochemical polymorphisms—detection, distribution, chromosomal location, and applications. In: Morse HC III, ed. Origins of Inbred Mice. New York: Acad Press, 1978:235-254.

17. Godfrey J, Searle AG. A search for histocompatibility differences between irradiated sublines of inbred mice. Genet Res 1965; 4:21-29.

18. Bailey DW. Mosaic histocompatibility of skin grafts from female mice. Science 1963; 141:631-633.

19. Bailey DW, Kohn HI. Inherited histocompatibility changes in progeny of irradiated and unirradiated inbred mice. Genet Res 1965; 6:330-340.

20. Bailey DW. Genetics of histocompatibility in mice I. New loci and congenic lines. Immunogenetics 1975; 2:249-256.

21. Snell GD. Congenic resistant strains of mice. In: Morse HC III, ed. Origins of Inbred Mice. New York: Acad Press, 1978:119-155.

22. Bailey DW. Recombinant-inbred strains: An aid to finding identity, linkage, and function of histocompatibility and other genes. Transplantation 1971; 11:325-327.

23. Bailey DW. Recombinant inbred strains and bilineal congenic strains. In: Foster HL, Small JD, Fox JD, eds. The Mouse in Biomedical Research. Vol I. New York: Acad Press, 1981:223-239.

24. Bailey DW. Histoincompatibility associated with the X chromosome in mice. Transplantation 1962; 1:70-74.

25. Lyon MF. Gene action in the X-chromosome of the mouse. Nature 1961; 190:372-373.

26. Strong LC. Transplantation studies on tumors arising spontaneously in heterozygous individuals. J Cancer Res 1929; 13:103-115.

27. Bailey DW. Heritable histocompatibility changes: Lysogeny in mice? Transplantation 1966; 4:482-487.

28. Kohn HI, Melvold RW, Dunn GR. Failure of x-rays to mutate class II histocompatibility loci in Balb/c mouse spermatogonia. Mut Res 1976; 37:237-244.

29. Kohn HI. H-gene (histocompatibility) mutations induced by triethylene melamine in the mouse. Mut Res 1973; 20:235-242.

30. Kohn HI, Melvold RW. Spontaneous histocompatibility mutations detected by dermal grafts: Significant changes in rate over a 10-year period in the mouse H system. Mut Res 1974; 24:163-169.
31. Bailey DW. How pure are inbred strains of mice? Immunology Today 1982; 3:210-214.
32. Hildemann WH, Cooper EL. Transplantation genetics: Unexpected histocompatibility associated with skin grafts from F2 and F3 hybrid donors to F1 hybrid recipients. Transplantation 1969; 5:707-720.
33. Bailey DW, Mobraaten LE. Histocompatibility of skin grafts from mice of F1, F2, and F3 generations on F1 generation hosts. Transplantation 1969; 7:567-569.
34. Bailey DW. Four approaches to estimating number of histocompatibility loci. Transpl Proc 1970; 2:32-38.
35. Gilmour DG. Numbers of genes influencing histocompatibility. Transplantation 1964; 2:426-428.
36. Bailey DW, Mobraaten LE. Estimates of the number of loci contributing to the histoincompatibility between C57BL/6 and BALB/c strains of mice. Transplantation 1969; 7:394-400.
37. Graff RJ, Brown DH. Estimates of histocompatibility differences between inbred mouse strains. Immunogenetics 1978; 7:367-373.
38. Peters J, Selley RL, Cocking Y. Mouse gene list 1997. Mouse Genome 1997; 95:193-466.
39. Roopenian DC. What are minor histocompatibility loci? A new look at an old question. Immunology Today 1992; 13:7-10.
40. Graff RJ, Silvers WK, Billingham RE et al. The cumulative effect of histocompatibility antigens. Transplantation 1966; 4:605-617.
41. Graff RJ. Minor histocompatibility genes and their antigens. In: Morse HC III, ed. Origins of Inbred Mice. New York: Acad Press, 1978:371-389.
42. Wettstein P. Immunodominance in the T-cell response to multiple non-H2 histocompatibility antigens. II. Observation of a hierarchy among dominant antigens. Immunogenetics 1986; 24:24-31.
43. Bailey DW. Genetically modified survival time of grafts from mice bearing X-linked histoincompatibility. Transplantation 1963; 2:203-206.
44. Wachtel SS, Gasser DL, Silvers WK. An association between H2 and the expressivity of H-Y in murine skin grafts. Transpl Proc 1973; 5:295-298.
45. Kralova J, Demant P. Expression of the H-y antigen on thymus cells and skin differential genetic control linked to K end of H2. Immunogenetics 1976; 3:583-594.
46. Melvold RW, Kohn HK. H2 and non-H2 interaction in expression of mutant histocompatibility gene H(KH-11). Immunogenetics 1977; 5:351-356.
47. Roopenian DC, Davis AP, Christianson GJ et al. The functional basis of minor histocompatibility loci. J Immunol 1993; 151:4595-4605.
48. Steinmuller D, Marcus JL, Bailey DW. The screening of non-H2 congenic lines for Sk loci. Immunogenetics 1978; 7:239-245.
49. Johnson LL, Bailey DW, Mobraaten LE. Genetics of histocompatibility in mice IV Detection of certain minor (non-H2) H antigens in selected organs by the popliteal node test. Immunogenetics 1981; 14:63-71.
50. Bailey DW. The vastness and organization of the murine histocompatibility-gene system as inferred from mutational data. In:Dausset J, Hamberger J, Mathe G, eds. Advance in Transplantation. Copenhagen: Munksgaard, 1968:317-323.
51. Moriwaki K. Wild mice from a geneticist's viewpoint. In: Moriwaki K, Shiroishi T, Yonekawa H, eds. Genetics in Wild Mice. Tokyo: Japan Scientific Societies Press, 1994:xiii-xxv.
52. Graff RJ, Bailey DW. The non-H2 histocompatibility loci and their antigens. Transplant Rev 1973; 15:26 49.
53. Rammensee HG, Klein J. Polymorphism of minor histocompatibility genes in wild mice. Immunogenetics 1983; 17:637-647.

CHAPTER 3

Some Personal Reflections on the History of Minor Histocompatibility Research

Hans-Georg Rammensee

Based on an article in Behring Institute Mitteilungen 94:11-16 (1994) and on a lecture held on the occasion of the Avery-Landsteiner-Award in Mainz, October 29, 1992.

The Start: Minor H Antigens as Models for Tumor Antigens

My first research project in immunology was given to me in 1979 by Jan Klein at the Max Planck Institute for Biology in Tübingen. He explained to me what minor histocompatibility (H) antigens are, or more exactly, what was known about them, and suggested that I test the polymorphism of minor *H* genes in wild mice. He also offered an alternative project but I can no longer remember its subject. I immediately agreed to work on minor H antigens, although I was not truly interested in the kind of H1, H3, or H4 antigens expressed by mice in the farm houses of Hohenentringen, Scotland or Egypt.[1] But the project required the production of cytotoxic T lymphocytes against minor H antigens, and I recognized that CTL against such comparatively weak antigens might have something in common with my real interest, which was tumor immunology. Now, almost 20 years later, I must say that I was not so very far away from correct idea, even as a young student.

The project dragged on, since the wild mice trapped at the farm houses had to be mated with laboratory mice to produce F_1 mice combining the wild minor H antigens with the required MHC restriction element, $H2^b$, for the CTL. The breeding required a lot of time which I used to think of ways of identifying the molecular nature of these antigens, which were at this point only known as genetic traits. As documented in my *Diplomarbeit* and also in my *Doktorarbeit*, both stored at the University of Tübingen, these trials failed miserably. For example, like several others before me, I made numerous attempts to raise polyclonal or monoclonal antibodies against minor H antigens, as is recorded in Table 3.1, which is translated from my thesis 'Regulation and effector phase of the immune response against minor histocompatibility antigens' (in German).[2] The only positive results were monoclonal antibodies against β2-microglobulin; however, since such monoclonals had been described already, the hybridomas were discarded. My quest to find out what molecules those minor H antigens might be was nevertheless fueled by this frustration. Several years later, in Mike Bevan's laboratory at Scripps in La Jolla, and subsequently at the Basel Institute for Immunology, I indeed managed to identify one minor H molecule from the H3 region as, again, being β2-microglobulin.[3] This, however, was an exceptional situation, because β2-microglobulin is the light chain of the MHC molecule.

Minor Histocompatibility Antigens: From the Laboratory to the Clinic, edited by Derry Roopenian and Elizabeth Simpson. ©2000 Landes Bioscience.

Table 3.1. Experiments carried out to obtain antibodies against minor H antigens[a]

Experiment	Strain Combination	Minor H Disparity	Immunization[b]
1	B10.LP-$H3^b$ (LPa) α B10	$H3^{b\alpha a}$	Skin graft (tested after rejection)
2	B10.129(5M)-$H1^b$ α B10	$H1^{b\alpha c}$	Skin graft (tested after rejection)
3	LPa α B10	$H3^{b\alpha a}$	Intraperitoneal (i.p.) injection of 2×10^7 glutaraldehyde-fixed spleen-, thymus- & lymph node cells[c]
4	LPa α B10	$H3^{b\alpha a}$	5 x i.p. injections of 2×10^7 spleen-, thymus- & lymph node cells
5	B6 α LPa	$H3^{a\alpha b}$	Skin graft after xenoimmunization with rat skin
6	LPa α B10	$H3^{b\alpha a}$	3 x i.p. injection of 2×10^7 spleen-, thymus- & lymph node cells together with T-cell growth factors into thymectomized, lethally irradiated mice reconstituted with syngeneic bone marrow
7	5M α LPa	$H1^{b\alpha c}$ $H3^{a\alpha b}$	8 x i.p. injection of 10^7 glass-adherent spleen cells
8	B10 α B10.129(21M)-$H4^b$	$H4^{a\alpha b}$	8 x i.p. injection of 10^7 Con A-blasts, LPS-blasts and epidermal cells[d]
9	B10.C-$H3^c$ α 21M	$H3^{c\alpha a}$ $H4^{a\alpha b}$	11 x i.p. injection of 10^7 anti-Thy-1 + complement-treated spleen cells
	B10.C-$H3^c$ α 21M	$H3^{c\alpha a}$ $H4^{a\alpha b}$	Same as above, but inoculum irradiated at 3,300 R

Table 3.1. Experiments carried out to obtain antibodies against minor H Antigens[a] (con't)

Experiment	Strain Combination	Minor H Disparity	Immunization
10	B10 α 21M	H4^{aab}	11 x i.p. injection of 10^7 anti-Thy-1 + complement-treated spleen cells into thymectomized, lethally irradiated mice reconstituted with syngeneic bone marrow and anti-Lyt-2.2 + complement-treated spleen cells
	B10 α 21M	H4^{aab}	Same as above, but inoculum irradiated at 3,300 R
11	LPa α B10	H3^{aab}	2 x subcutaneous injection of 2×10^7 spleen cells in complete Freund's adjuvant; subsequently fusion of cells from the drained lymph nodes with NS-1 myeloma cells.[e] Hybridoma supernatants were tested for antibodies
12	5M α LPa	H1^{bac}, H3^{aab}	6 x i.p. injection of 10^7 spleen cells. Cell-free lysate from spleen cells was tested for antibodies

a The serum (cell lysate in experiment 12; hybridoma supernatants in experiment 11) was tested cytotoxically[23] and in some experiments by indirect fluorescence staining[24] or CML inhibition.[25] All the sera were found to be negative.

b Blood was tested seven days after the final immunization (except for experiments 11 and 12).

c Method according to Sanderson and Frost.[26]

d Method according to Long et al.[27]

e Method according to Kohler and Milstein.[28]

The Hypothesis: Minor H Antigens are Short Processed Peptides

Just at the time the paper describing the H3/ β2-microglobulin work went to the printer, I became aware of Alan Townsend's hypothesis on the nature of minor H antigens:[3] He proposed that they are fragments, or peptides, of cellular proteins and that these peptides were presented by MHC class I molecules, just like viral antigens. This appealing theory made me consider possibilities for isolating such hypothetical fragments from minor H antigen expressing cells by biochemical methods. However, I was no expert biochemist. The most sophisticated path I could take was to purify antibodies over a protein A column. I discussed the problem with a number of people in Basel and elsewhere and all MHC experts were of the opinion that my idea was madness. From discussions with John Lambris, I learned that for the isolation of peptides from a complex mixture, as was to be expected, reversed phase HPLC would be a reasonable and probably the only possible method. Thus, I decided in 1986 that an HPLC system would have to be purchased. This was in 1986. Unfortunately, HPLC systems were terribly expensive. I applied for one in my new laboratory at the Max Planck Institute in Tübingen where I moved in 1987, which was the same year Pamela Bjorkman, Don Wiley and colleagues described the first crystal structure of an MHC class I molecule, suggesting that MHC molecules were peptide binding molecules.[5] However, the Max Planck Institute had already agreed to buy a FACScan machine for my laboratory and there was no possibility of ordering another piece of expensive equipment, so the enormous sum of DM 70000—for the HPLC—had to be funded from grant money. Having been out of the country for several years, I was not very familiar with the German grant system, which caused some delay. Upon sending a grant proposal to the Deutsche Forschungsgemeinschaft in Bonn, I was advised to seek membership in one of the local "Sonderforschungsbereiche" (SFB 120) that are also funded by the DFG, since the project was related to its topic: 'Leukemia research and immunogenetics.' My project proposal for the local SFB commission must have been convincing because in February 1988, the SFB, headed by Professor Waller, decided to accept my application for membership. No money was immediately available and I had to wait until 1989 before the next three-year term of the SFB began. Before that, I had to put forward my proposal to the national committee evaluating the SFB research plan for 1989-1991. The committee approved my membership with the SFB, although my grant was reduced. I was not granted the postdoctoral position that I had applied for but more importantly, the HPLC was approved. On the April 4, 1989, the HPLC arrived in our lab.

I am thankful to the DFG for many things in my scientific life. It had financed my postdoctoral fellowship with Mike Bevan and prolonged my membership in the SFB for a further 3 years. In 1992, I received an astronomically large grant in the form of the Leibniz Prize. The most important thing, however, was the DFG's approval to purchase the HPLC.

The installation of the HPLC was accompanied by very mixed feelings. We had spent the enormous sum of DM 73000—from the taxpayer's pocket and I did not even know if it would be of any use for our purposes. All the experts had warned me. Nevertheless, four weeks later a student in the lab, Hajo Wallny, managed to isolate a peptide representing a minor H antigen by HPLC.[6] He prepared a protein extract from BALB.B spleen cells, digested it with proteases, separated the resulting fragments on the HPLC, and tested the individual fractions for recognition by minor H-specific CTL (B6 against BALB.B). Thus, Townsend's hypothesis on minor H antigens as fragments of normal cellular proteins was confirmed. But these peptides were digested artificially, with enzymes from the biochemistry catalogue. How would the peptides produced by the cell itself look like?

This problem was solved by two undergraduate biochemistry students, Olaf Rötzschke, who came to my lab to carry out a practical course (Praktikum) in the summer of 1988, and Kirsten Falk, who came shortly afterwards. The first attempt to solve the problem was made by trying to immunoprecipitate MHC class I molecules from cells and to dissociate the postulated peptides, as described by Soren Buus, Stephane Demotz, Howard Grey and colleagues for class II molecules.[7] For some reason, however, the immunoprecipitation of MHC molecules did not work; there was never a band on the gel with the expected size. Frustrated by this experience, one of the two had the ingenious idea for a simple, straightforward experiment: cells were disrupted in the starting buffer for reversed phase HPLC, 0.1% trifluoroacetic acid (TFA). It was reasoned that this treatment should dissociate peptides noncovalently bound to other molecules and leave them in solution. Thus, the supernatant should contain a complex mixture of cellular material that is soluble in 0.1% TFA, but including peptides that were bound to MHC molecules. This experiment was first carried out with H2^d-expressing mouse tumor cells transfected with *E. coli* β-galactosidase. The TFA-extract of these cells was separated on HPLC and the individual fractions were then tested for recognition by a β-galactosidase specific, H2-L^d-restricted CTL line.[8] When I first saw the results it looked too good to be true. The HPLC profile was extremely complex, as expected. But the CTL recognized just one single fraction, and this recognition was very strong even when the fraction was diluted. Thus, I tried to remain skeptical, and insisted that the same approach should be tried for other antigens, for the minor H antigens HY and H4. Kirsten and Olaf managed these experiments very quickly and successfully.[8] Thus, we had a simple method to isolate naturally processed MHC class I restricted T-cell epitopes from cells. This started a tremendously productive and exciting period, in which Kirsten and Olaf worked every day from late morning to midnight or later, thereby postponing their preparation for their final exam in biochemistry.

The material eluted by acid extraction and recognized by CTL was indeed of peptidic nature, since it could be destroyed by proteases. An unexpected finding was that these peptides were dependent on the MHC molecules of the cells from which they were isolated. The HY peptide, for example, which is recognized by H2-D^b-restricted CTL, could not be isolated from male BALB/c cells, or any other male cells not expressing D^b.[9] Thus, this observation not only strongly suggested that the peptides had indeed been MHC-associated in the cell, they also indicated that MHC molecules must be involved in the production or in the protection of its ligands. Confirmation that these peptides were indeed MHC ligands was obtained by applying the acid extraction method to immunoprecipitated MHC molecules, which we finally managed to carry out.[10] The acid extraction approach was then quickly used for a number of other antigens, especially viral antigens and those recognized by alloreactive T cells.[10,11] By comparing peptides isolated from influenza-infected cells with synthetic peptides, in collaboration with Günther Jung's group at the University of Tübingen, it was possible to identify the naturally presented K^d-restricted peptide from influenza nucleoprotein to be TYQRTRALV, and the D^b-restricted peptide from the same protein to be ASNENMETM.[11,12] A number of synthetic peptides had already been described at that time that bind to K^d, and Tyr was important for binding, as demonstrated by J. Maryanski and colleagues.[13] Comparison of the K^d-restricted TYQRTRALV with these synthetic K^d-binders suggested to us that natural K^d ligands might be nonamers with Tyr at position 2 and an aliphatic residue at the C-terminus. These features seemed to be specific for K^d, since neither the D^b-restricted nucleoprotein peptide nor a K^b-binding ligand from vesicular stomatitis virus described at approximately the same time by G. van Bleek and S. Nathenson[14] actually met these criteria.

MHC-Binding Motifs

Since Kirsten and Olaf had just managed to immunoprecipitate MHC molecules, I suggested precipitating a large amount of K^d molecules, isolating the peptides, and sequencing the native peptide mixture together in order to test the hypothesis. Position 2 of such pool

Table 3.2. The K^d-restricted peptide motif[a]

	Position								
	1	2	3	4	5	6	7	8	9
Dominant anchor residues		Y							I L
Other residues	K A R S V T	F	N D L A H V R	P A E S D H N	M V N D I L S T G	K F H I M Y V R L	T N P H D E Q S	H E K V V F R	
Weak			S F E Q K M T						
Examples for ligands	T S	Y Y	Q F	R P	T E	R I	A T	L H	V I
Protein source									Protein tyrosine kinase JAK1

a. Large numbers of mouse tumor cells expressing K^d (P815) were lysed with detergent. K^d molecules were immunoprecipitated and treated with TFA to dissociate peptides. The resulting supernatant was separated by HPLC. Fractions thought to contain K^d ligands were pooled and sequenced by Edman degradation. The table indicates the amino acids (single letter code) detected in the different cycles of Edman degradation (reflecting the positions within the peptide). For example, at position 2, only the aromatic residues Tyr and Phe were found, with Tyr far more abundant that Phe. Positions 3 through 8 contained a number of different amino acids. At position 9, however, only aliphatic residues were found, and none at positions 10 or beyond. The natural K^d ligand from Influenza nucleoprotein, TYQRTRALV fits well to the motif, since Val is an aliphatic residue like Ile or Leu. The first MHC-ligand to be directly sequenced, SYFPEITHI, also fits well to the motif.[15,16]

Table 3.3. *Peptide motifs of some frequent HLA molecules*[a]

	Position									
	1	2	3	4	5	6	7	8	9	
HLA-A1	-	-	D E	-	-	-	-	-	Y	
Example for ligands	E	A	D	P	T	G	H	S	Y	Melanoma-associated antigen MAGE1
	A	T	D	F	K	F	A	M	Y	Normal cellular protein
HLA-A2	-	L M	.	-	-	-	-	-	V	
Examples for ligands	G	-	L	G	F	V	F	T	L	Influenza matrixprotein
	-	L	K	E	P	V	H	G	V	HIV reverse transcriptase
	S	L	L	P	A	I	V	E	L	Normal cellular protein
HLA-B8	-	-	K	-	K R	-	-	-	L	
Examples for ligands	G	P	K	V	K	Q	W	P	L	HIV reverse transcriptase

[a] These peptide motifs describe the features required by the individual MHC molecules for presentation. For A2, for example, the peptides consist of 9 amino acids and have Leu, Met or Ile at position 2 and Val or Leu at P9, as also seen in the examples. For B8, the ligands are nonamers as well; however, they must have positively charged residues at P3 and P5 (Lys or Arg) and an aliphatic residue such as Leu at P9. Exceptions and further details are not considered in this table.[15,16,21]

sequencing should stand out as a Tyr signal and after position 9, the signals should drop. In addition, the problem of the small amounts of individual MHC ligands which were too low for sequencing by routine methods could be circumvented this way. Luckily, our peptide chemist colleagues were open-minded and agreed to carry out this experiment: Stefan Stevanovia Ph.D. student in Günther Jung's group, actually did the first pool sequencing on October 30, 1990. The results confirmed the hypothesis, and indicated a very clear motif (Table 3.2).[15] The same approach was quickly applied to other class I molecules: D^b, K^b, D^d, and L^d. For D^b and K^b, clear motifs could be seen which were different from those of K^d and fitted well to the only ligands that were known. For D^d and L^d, the amounts of peptides were lower, but motifs could be recognized as well. Convinced that we had found something really important, we sent a paper describing the K^d, K^b, and D^b motifs to *Science* just before Christmas. To our disappointment, the paper was returned to us in January without even being read by reviewers. *Nature* did send the manuscript to reviewers and both were fairly enthusiastic; nevertheless, one insisted that we include the HLA-A2 motif in our study as well, since the crystal structure of this molecule was known. In the meantime we had done exactly that and could actually include the data immediately.

Conclusion

Looking at these motifs and the ones described thereafter[16] one can summarize their features as follows: MHC class I molecules are peptide receptors with a peculiar specificity. Each allelic product has its individual specificity, characterized by peptide length, usually 8 or 9 amino acids, and one or two anchor positions that have to be occupied by only one or one of a few similar residues (Table 3.3). One of these anchors is always at the C-terminus, and always either hydrophobic or positively charged. The peptide specificity of MHC class I molecules can now be explained in depth by more detailed crystallographic studies.[17-20]

In the meantime, other groups (see other chapters in this book) have identified quite a number of minor H peptides, by a combination of genetic and peptide elution approaches. This allowed us to include a small list of minor H peptides into our 1997 list of MHC-associated peptides,[21] in addition to the much larger tables on viral or self peptides.

Looking back on our research, it is evident that the original quest to identify a few exotic minor H antigens was only partially successful but led to the discovery of the principle of the allele-specific peptide receptor function of MHC molecules, shed light on the centerpiece of the specific immune system, and was of major importance for applied immunology and even tumor immunology.

References

1. Rammensee H-G, Klein J. Polymorphism of minor histocompatibility genes in wild mice. Immunogenetics 1983; 17:637-647.
2. Rammensee H-G. Regulation und Effektorphase der Immunantwort gegen schwache Histokompatibilitätsantigene. Dissertation, University of Tübingen, 1982.
3. Rammensee H-G, Robinson PJ, Crisanti A et al. Restricted recognition of beta 2-microglobulin by cytotoxic T lymphocytes. Nature 1986; 319:502-504.
4. Townsend AR, Gotch FM, Davey J. Cytotoxic T cells recognize fragments of the influenza nucleoprotein. Cell 1985; 42:457-467.
5. Bjorkman PJ, Saper MA, Samraoui B et al. Structure of the human class I histocompatibillty antigen, HLA-A2. Nature 1987; 329:506-512.
6. Wallny H-J, Rammensee H-G. Identification of classical minor histocompatibillty antigen as cell-derived peptide. Nature 1990; 343:275-278.
7. Grey HM, Demotz S, Buus S et al. Studies on the nature of physiologically processed antigen and on the conformation of peptides required for interaction with MHC. Cold Spring Harbor Symp Quant Biol 1989; 54:393-400.

8. Rötzschke O, Falk K, Wallny H-J et al. Characterization of naturally occuring minor histocompatibility peptides including H-4 and H-Y. Science 1990; 249:283-287.
9. Falk K, Rötzschke O, Rammensee H-G. Cellular peptide composition governed by major histocompatibillty complex class I molecules. Nature 1990; 348:248-251.
10. Rötzschke O, Falk K, Faath S, Rammensee H-G. On the nature of peptides involved in T cell alloreactivity. J Exp Med 1991; 174:1059-1071.
11. Rötzschke O, Falk K, Deres K et al. Isolation and analysis of naturally processed viral peptides as recognized by cytotoxic T cells. Nature 1990; 348:252-254.
12. Falk K, Rötzschke O, Deres K et al. Identification of naturally processed viral nonapeptides allows their quantification in infected cells and suggests an allele-specific T-cell epitope forecast. J Exp Med 1991; 174:425-434.
13. Maryanski JL, Verdini AS, Weber PC et al. Competitor analogs for defined T-cell antigens: Peptides incorporating a putative binding motif and polyproline or polyglycine spacers. Cell 1990; 60: 63-72.
14. Van Bleek GM, Nathenson SG. Isolation of an immunodominant viral peptide from the class I H-2K^b molecule. Nature 1990; 348:213-216.
15. Falk K, Rötzschke O, Stevanovi S et al. Allele-specific motifs revealed by sequencing of self-peptides eluted from MHC molecules. Nature 1991; 351:290-296.
16. Rammensee H-G, Falk K, Rötzschke O. Peptides naturally presented by MHC class I molecules. Annu Rev Immunol 1993; 11:213-244.
17. Madden DR, Gorga JC, Strominger JL et al. The three-dimensional structure of HLA-B27 at 2.1 A resolution suggests a general mechanism for tight peptide binding to MHC. Cell 1992; 70;1035-1048.
18. Madden DR, Garboczi DN, Wiley DC. The antigenic identity of peptide-MHC complexes— a comparison of the conformations of five viral peptides presented by HLA-A2. Cell 1993; 75:693-708.
19. Zhang W, Young ACM, Imarai M et al. Crystal structure of the major histocompatibility complex class I H-2K^b molecule containing a single viral peptide: Implications for peptide binding and T-cell receptor recognition. Proc Natl Acad Sci USA 1992; 69:8403-8407.
20. Fremont DH, Matsamura M, Stura EA et al. (1992) Crystal structures of two viral peptides in complex with murine MHC class I H-2K^b. Science 1992; 257:919-92Z
21. Rammensee H-G, Bachmann J, Stevanovi S. MHC ligands and peptide motifs. Austin: Landes Bioscience. New York: Springer-Verlag, 1997.
22. Falk K, Rötzschke O. Consensus motifs and peptide ligands of MHC class I molecules. Sem Immunol 1993; 5:81-94.
23. Klein J, Hauptfeld V, Hauptfeld M. Evidence for a fifth region (G) in the H-2 complex of the mouse. Immunogenetics 1975; 2:141.
24. Geib RW, Klein J. MLR blast cells generated in mutant-standard strain combinations bind H-2K and H-2D antigens. Eur J Immunol 1979; 9:135.
25. Koo CG, Varano A. Inhibition of H-Y cell-mediated cytolysis by monoclonal H-Y-specific antibody. Immunogenetics 1981; 14:183.
26. Sanderson CJ, Frost P. The induction of tumour immunity in mice using glutaraldehyde-treated tumour cells. Nature 1974; 248:690.
27. Long PM, Lafuse WP, David CS. Serologic and biochemical identification of minor histocompatibility (H-4) antigens. J Immunol 1981; 127:825-828.
28. Köhler G, Milstein D. Continuous cultures of fused cells secreting antibody of predefined specificity. Nature 1975; 256:495.

CHAPTER 4

Identifying T Cell-Defined Histocompatibility Antigens by Expression Cloning

Lisa M. Mendoza, Pedro Paz, Aamir Zuberi, Gregory Christianson, Derry Roopenian, and Nilabh Shastri

Minor histocompatibility (H) antigens were first defined at the cellular level approximately 50 years ago.[1,2] Since then, over 50 distinct H loci have been identified based upon their inheritance and segregation patterns in recombinant and congenic mouse strains. Polymorphisms in the H antigens account for the immunological responses that lead to graft rejection and graft versus host disease (GVHD) when tissue is transplanted between MHC identical individuals.[3] The gene products of H loci serve as precursors of processed peptides which are presented by MHC molecules to T cells.[4] The graft rejection phenotype resulting from differences between H loci, therefore, is the result of T-cell responses elicited by either qualitative or quantitative differences in the antigenic peptides displayed by the MHC molecules on the donor versus the host cell surface. Identification of H antigens at the molecular level is thus critical for defining the basis of these genetic polymorphisms and is important for understanding the mechanisms by which graft rejection occurs.

H antigens may arise from a variety of potential differences in the genetic backgrounds of MHC-matched individuals. These include differential H gene transcription and expression of the protein product which may affect the presence of the protein precursor; the antigen processing pathway where differences may affect the generation of the antigenic peptide; nucleotide substitutions within the antigen gene sequence which result in amino acid substitutions within the antigenic peptide and which could affect either the ability of the peptide to bind the MHC molecule or the ability of the T-cell receptor (TCR) to recognize the presented peptide. In addition to the effects of these polymorphisms on transplantation, H antigenic differences may have direct relevance to selection of the T-cell repertoire in the thymus and in the periphery (Fig. 4.1). H antigens resulting from transcriptional polymorphisms which affect either expression levels or mRNA stability have been identified recently.[5] Analogous to many tumor antigens, differential expression of the antigenic precursor could result in lower levels of peptide than can be detected by the TCR. In this case, the polymorphism would be due to levels of presented peptide below the threshold required for negative selection. Additionally, differences in regulatory elements could affect the tissue distribution of the antigenic precursor which might result in inefficient presentation of the antigen, for example, if it is presented by a nonprofessional antigen presenting cell (APC). Polymorphisms resulting from differences in the translational machinery of the cell affecting either expression levels or the sequence of the antigenic peptide is affected might also result in H antigens similar to those described above.

Minor Histocompatibility Antigens: From the Laboratory to the Clinic, edited by Derry Roopenian and Elizabeth Simpson. ©2000 Landes Bioscience.

Mechanisms for immunogenicity of minor histocompatibility antigens

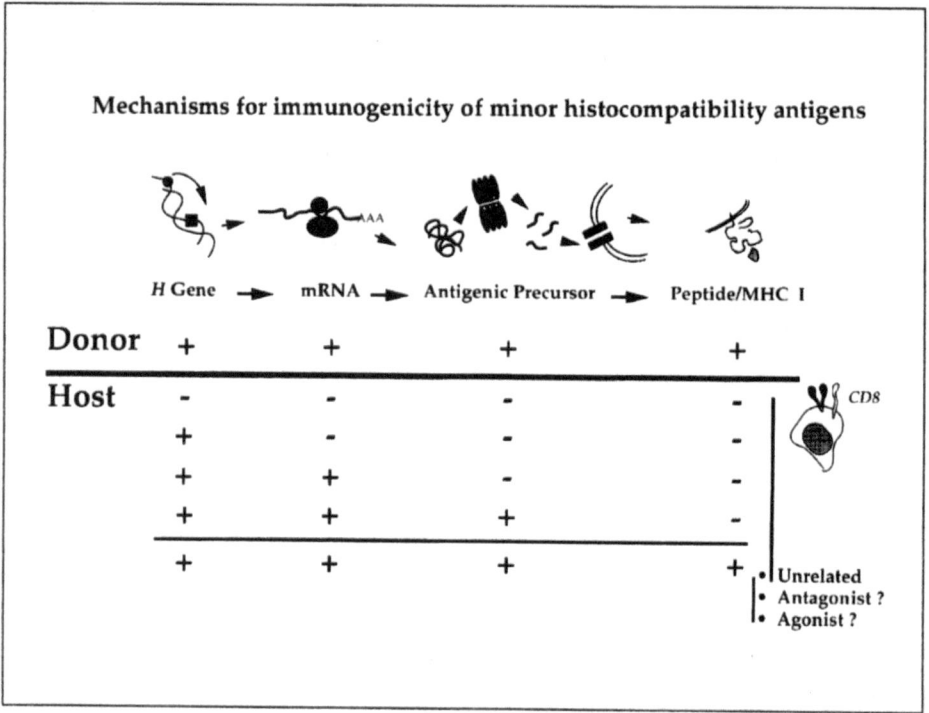

Fig. 4.1. Mechanisms for immunogenicity of minor histocompatibility antigens. Polymorphisms at the level of replication, transcription or translation of the protein precursor between the donor and host strains can give rise to H antigens. In addition, differences in the antigen processing pathway and generation of the processed peptide that affect either binding of the peptide to the MHC Class I molecule or recognition of the peptide/MHC complex by the T-cell receptor can result in H antigens.

Polymorphisms in the antigen processing machinery, for example differences in protease specificities or in the TAP requirements, might also lead to the generation of H antigens. Sequence polymorphisms in the antigenic precursor which affect processing of the antigenic peptide could also give rise to H peptides. It has been shown that differences in residues that flank the antigenic peptide can affect the efficiency by which the peptide, in both minigenes and in the full length protein precursor, is presented to T cells.[6,7] Sequence polymorphisms within the antigenic peptide that affect the ability of the peptide to bind MHC would result in an H antigen that is essentially foreign,[8-10] whereas polymorphisms that accommodate MHC binding but affect TCR recognition may potentially act as weak agonists or may even act as antagonists during T-cell selection in the thymus, although these possibilities have not been tested directly.[11,12] These are some of the many possible scenarios that could result in H antigenic responses and until a significant number of these peptides and the genes that encode them are identified at the molecular level we can only speculate on the molecular basis of their polymorphic phenotype.

Minor Histocompatibility Antigens Identified at the Molecular Level

To date, six nuclear and three mitochondrial proteins containing H antigens are known where the gene, the protein precursor and the antigenic peptide have been defined (Table 4.1).[13] This includes the mitochondrially encoded H antigens ND1, COI and ATPase 6,[11,14,15] Y chromosome-linked H antigens derived from *Smcy* and *Uty*,[10,12,16,17] the first autosomal H antigen H13,[12] reviewed in detail here, and three very recently identified autosomal H antigens, the human HA-1,[18] and the mouse H60[5] and H3a[19] antigens. Identification of such H peptides has been a long time coming due to the lack of efficient and practical methodologies to identify T-cell peptide epitopes. Three basic technical strategies have been used to identify antigenic peptides:

1. positional cloning, which requires a knowledge of the chromosomal position of the gene responsible for the H antigen;
2. biochemical purification, which requires sophisticated instrumentation, experimental and analytical skills; and
3. expression cloning, which requires construction of representative cDNA libraries, the efficient expression of the antigen gene and presentation of the appropriate peptide/MHC complex on the APC.[20]

All three approaches have proven successful in the identification of H antigens.

Positional cloning approaches exploit chromosomal location of the *H* gene as a starting point. It was possible to employ a positional cloning approach to identify the genes responsible for mouse maternally transmitted antigens (Mta) because they mapped to the mitochondrial genome.[11,21] The available sequence information and the relatively small size of the mitochondrial genome (10-30 Kb) facilitated this approach. The 13 proteins encoded by the mitochondrial genome have been previously defined and maternally transmitted antigens (Mta) are now genetically well characterized.[22,23] Two of these antigens are N-formylated peptides presented by the H2-M3 class I molecule (Chapter 5).[15,16] These properties facilitated their identification since the search for the antigenic peptide was limited to the amino terminal sequence of the 13 mitochondrial proteins. The identification of three mouse Y chromosome-encoded HY antigens, one presented by the K[k] and the other two by the D[b] class I molecule, has also been greatly facilitated by the genetic position and the limited complexity of the Y chromosome compared with the autosomes (Chapter 6).[8,10]

In the case of autosomally encoded H antigens, positional cloning is a much more daunting task. Genetic positional information is available with a low degree of resolution for many of the 50 distinct autosomal mouse minor *H* loci that have been functionally mapped to a particular chromosomal location, largely by the use of congenic mouse strains,[24] however, comparable information is more difficult to ascertain in humans. Even the most refined conventional genetic analysis rarely yields resolution less than several megabases of genomic DNA, which could contain hundreds of genes, all of which are potential candidates. Recently, however, positional cloning has resulted in the identification of the gene encoding the autosomal mouse *H* gene, *H3a* (Chapter 1).[19] Finally, although the *H* gene might functionally map to a particular chromosome based on T-cell reactivity, the possibility that the structural gene is present on another chromosome and is regulated in trans cannot be ruled out.

Identification of antigenic peptides by the biochemical purification approach requires a large amount of starting material ($\sim 10^{12}$ cells expressing the antigen) depending on the abundance of the peptide and the sensitivity of the T cell.[20] Furthermore, the identity of the antigenic precursor gene is not necessarily defined when the antigenic peptide is obtained by biochemical purification. This is a major limitation of this purification approach since identification of the gene requires that a match be found in the nucleotide or protein databases. This requirement has been satisfied for the HY peptide epitopes derived from the human

Table 4.1. Defined minor H antigen genes

Species	H Ag	Chromosomal Location	Protein Precursor	MHC Molecule	Antigenic Peptide [donor, host allele]	Complexes/cell	Tissue Expression	Mechanisms of Polymorphism	Ref.
Mouse	MTF	Mitochondria	ND1	H2-M3	fMFFIN[I,A]LTL	?	ubiquitous	TCRbiallelic	14
Mouse	-	Mitochondria	COI	H2-M3	fMF[I,T]NRW	?	ubiquitous	TCRrec	15
Rat	-	Mitochondria	ATPase6	RT1.A^a	ILFPSS[E,K]RLISNR	?	ubiquitous	TCRrec	11
Mouse	HY	Y	Smcy	H2-K^k	TENSGKDI	?	ubiquitous	Nopep	8
Mouse	HY	Y	Smcy	H2-D^b	KCSRNRQYL	?	ubiquitous	?	17
Human	HY	Y	SMCY	HLA-B7	SP[S,A]VDKA[R,Q]AEL	?	ubiquitous	TCRrec	9
Human	HY	Y	SMCY	HLA-A2	FIDSYIC*QV	?	ubiquitous	Nopep	16
Mouse	HY	Y	Uty	H2-D^b	WMHHIN,T[IM,V]DL[I,L]	?	ubiquitous	MHCbind	10
Mouse	H13	2	Novel protein	H2-D^b	SSV[V,I]GVWYL	~50	ubiquitous	TCRbiallelic	12
Human	HA-1	?	Novel protein	HLA-A*0201	VL[H,R]DDLLEA	?	?	MHCbind	18
Mouse	H60	10	Novel protein	H2-K^b	LTFNYRNL	~15	?	Expression	5
Mouse	H3a	2	2fp106	H2-D^b	[A,T]SP[C,R]NSTVL	~160	ubiquitous	TCRbiallelic	19

*Indicates cysteinylated cysteine residue required for T-cell recognition.
#MHCbind: amino acid change(s) affects binding of antigenic peptide to MHC molecule.
TCRrec: amino acid change(s) affects TCR recognition but not binding of the antigenic peptide to MHC molecule.
TCRbiallelic: biallelic TCR recognition where reciprocal immunogenicity has been demonstrated for both allelic forms of the antigen.
Nopep: no similar peptide in host X chromosome homologue.
Expression: gene expression is affected in one allelic strain resulting in a null allele.

SMCY protein[9,16] and the HA-1 antigen[18] but the identity of the human HLA-A2.1 restricted HA-2 peptide[27] remains unknown due to the absence of any matches in the database. Without identification of the precursor protein allelic differences cannot be determined and consequently the basis for the polymorphism remains unclear. Despite this limitation, three H peptides have been identified by biochemical means. The advantages offered by this approach include the inherent proof that the T-cell stimulating peptide is the naturally processed antigen and the ability to identify peptides that contain post-translational modifications. These peptides may only be identifiable by direct sequencing of the naturally processed peptide as was shown recently for the cysteinylated HY peptide from the SMCY protein and the tyrosinase tumor antigen.[16,26]

The third approach applied to the identification of H antigens is expression cloning. This approach requires the generation of a representative cDna library from the tissue expressing the antigen or a complete genomic library and transfection of the library as pools of DNA into recipient APCs for expression, processing and presentation of the antigenic peptide. The autosomal H13 antigen has been identified by this approach.[12] More recently, we have also identified the H60, H47 and H28 antigens using a similar approach.[5,27] This method has also had extensive success in the search for tumor antigens.[28]

Expression cloning requires a smaller number ($\sim 10^{6-8}$) of cells as starting material to generate a representative cDna library. This is at least 10^4 times fewer cells than required for the purification approach and is an amount that can easily be obtained from organs or cells in culture. The methods for generating and propagating cDNA libraries are general molecular biology methods[29] and the expression vector that we and others have applied in this approach has been well characterized in several mammalian cell lines.[30] In addition, knowledge of the chromosomal location of the gene of interest is not a prerequisite for the success of this approach. Expression cloning is limited when searching for low abundance transcripts or antigens whose precursor protein is large. For example, an antigenic peptide contained within a transcript greater than 5 Kb might not be easily cloned if the peptide is present at the 5' end of the gene, as was the case for the positionally cloned *H3a* gene.[19] Cotranslational modifications of the antigenic peptide that are necessary for T-cell recognition may also be problematic if they do not occur in the recipient APC. Barring these limitations, confirmation of the antigenic activity as the naturally processed peptide might also prove difficult for post-translationally modified peptides. Identification of the naturally processed peptide requires coelution by HPLC of the antigenic activity identified in cell extracts with synthetic peptides which would not possess any modification. Lack of coelution by HPLC could, however, provide clues to the nature of such posttranslational modifications as was demonstrated for the tyrosinase antigen gene first identified by expression cloning.[26,31]

Expression Cloning: Identification of the H13 Antigen

We have developed a novel expression cloning strategy which employs the use of T-cell hybrids that express the reporter gene lacZ upon activation via the T-cell receptor (lacZ-inducible) and can be used in both bulk assays and at the single cell level to measure T-cell activation.[32] Using these lacZ-inducible T cells as probes for screening cDNA libraries we proposed a novel expression cloning approach for the identification of T-cell antigens several years ago.[33] This expression cloning approach has had success in our laboratory in the identification of several antigens recognized by allo- and tumor-specific CD8$^+$ T cells[34-36] and most recently in the identification of the H13 antigen.[12] Briefly, this method involves screening DNA pools from a representative cDna library transiently transfected into recipient APCs that are co-transfected with the restricting class I MHC cDNA and the costimulatory ligand B7-2 cDNA. The positive pool(s) is identified and the relevant cDNA cloned. Identity of the candidate as the antigenic precursor is confirmed by restriction to the appropriate MHC class I molecule. The antigenic peptide is then determined by deletional analysis using both minigene constructs and synthetic peptides. Identification of the naturally processed peptide is demonstrated by coelution of the T-cell stimulating activity from cell extracts with the synthetic peptide by HPLC. The molecular basis for the H polymorphism is then determined by isolation and sequencing allelic homologues by either RT-PCR or PCR from genomic DNA.

We applied this approach to the identification of the first autosomally encoded H antigen, H13.[12] An H13ª specific CTL clone was fused with the lacZ-inducible, TCRα-βWZ.36/CD8⁺ fusion partner [37] to generate T-cell hybrids specific for H13ª. The resulting hybrid, 30NX/ B10Z, expressed lacZ upon engagement of the TCR and could detect the H13ª antigen at levels comparable to the parent CTL clone. This hybrid was used as a probe to screen a cDna library constructed in the eukaryotic expression vector pcDNA1 (Invitrogen, La Jolla, CA). The cDNA was generated from an H13ª expressing thymoma cell line EL4, which presents H13ª to the 30NX/B10Z T cells.

The EL4 cDNA pools were simultaneously screened with several T-cell probes including 30NX/B10Z following transient transfection into recipient cells cotransfected with cDNAs encoding the restricting class I MHC molecule and the costimulatory ligand B7-2 using DEAE dextran and chloroquine in 10% NuSerum. The cells were allowed to recover and express the transfected DNAs for 48 h at which time the T cells were added to the transfected APCs. The T cells were then analyzed for expression of lacZ one day later in bulk assays (Fig. 4.2). When screening for the H13 antigen, we identified a putative positive and subdivided the positive pool of DNAs into individual colonies. Once cloned, the DNA was tested for its ability to stimulate the T cells in the presence of the restricting class I MHC and not in the presence of an irrelevant class I MHC molecule. This test confirmed that T-cell activation occurs via the TCR and not non-specifically when the cloned cDNA is transfected into the APC.

Once the identity of the cloned H13 cDNA was confirmed, the antigenic peptide was defined by deletional analysis. Deletion constructs from the 3' end of the cDNA were made by PCR using a vector-specific forward primer and cDNA-specific reverse primers. Nested deletions made in this way mapped the region containing the antigenic activity to within 108 bp. Once mapped, the minimal cDNA sequence was analyzed for putative Dᵇ MHC class I binding motifs[38] and minigene constructs within the region were tested for the ability to stimulate the H13ª-specific T cells. These experiments surprisingly showed that the H13ª antigenic peptide did not match the Dᵇ MHC motif (xxxxNxxx[I,L,M]) and required further mapping of the 108 bp region by additional PCR deletions to determine the minimal peptide required to stimulate the T cells. The minimal peptide, SSVVGVWYL, was confirmed by both endogenous and exogenous presentation assays and its identity as the naturally processed antigen was established by co-elution with the synthetic peptide by HPLC.

To define the molecular basis for the H13 antigen, we cloned a region of the gene from genomic DNA of both allelic strains of mice by PCR with H13-specific primers. We found that a single conservative amino acid substitution from valine to isoleucine at position 4 of the antigenic peptide resulted in a dramatic loss of T-cell activation in endogenous presentation assays. In exogenous presentation assays with synthetic peptides we observed a cross-reactivity of the 30NX/B10Z T cells to the self-peptide that required 500 fold more peptide. In reciprocal immunizations with the H13 congenic strains B10 (H13ª) and B10.CE (30NX)-*H13ᵇ* we showed that both H13ª- and H13ᵇ-specific T cells could be elicited and that they recognized the cognate peptides SSV[V,I]GVWYL (S[V,I]L9). The presence or absence of the single methyl group in the S[V,I]L9/Dᵇ complex provides the primary structural basis for the self/nonself discrimination by the T cell repertoire. Interestingly,

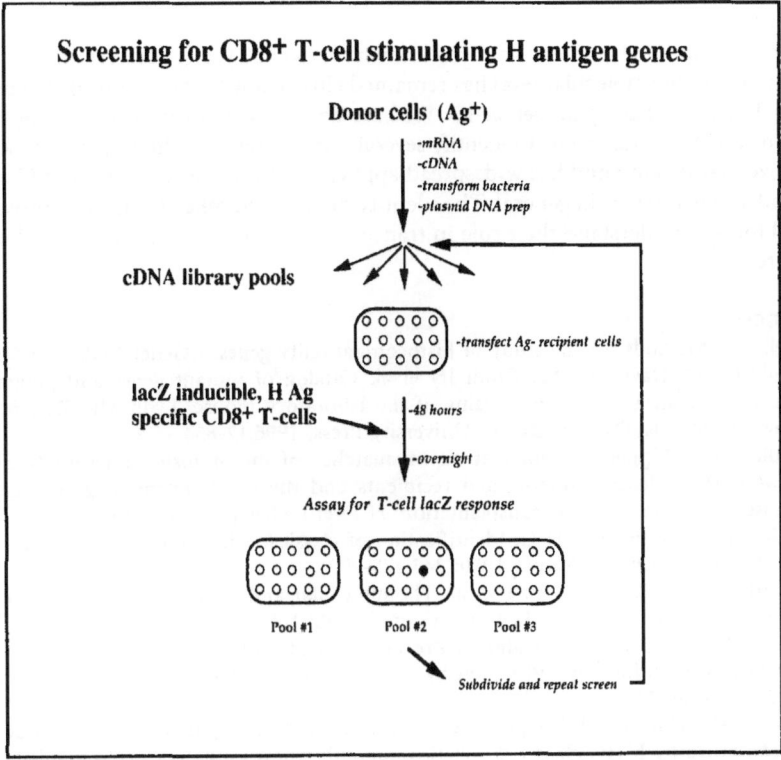

Fig. 4.2. Screening for CD8⁺ T-cell stimulating H antigen genes. A cDNA library from donor cells expressing the H antigen is divided into pools as plasmid DNA and transfected into recipient APCs. After 48h lacZ-inducible T cells are added to the transfected APCs and cultured overnight. LacZ response is measured to assay for T-cell activation. Positive pool(s) are subdivided and rescreened until the cDNA encoding the H gene is cloned.

similar observations were made which distinguished the mitochondrial H antigenic peptides ND1α and ND1β (I->A at position 6) presented by the H2-M3 class 1b MHC molecule[14] as well as the allelic ATPase6 peptides (E->K at position 7) presented by RT1.A^a, the rat MHC class I molecule (Table 4.1).[11] Unlike these H antigens, however, female mice fail to present X encoded homologues of the HY antigens from either Uty or Smcy due to the lack of homologous peptide and/or the lack of MHC-binding motifs (Table 4.1). With respect to these epitopes, the surface of male cells is thus unambiguously foreign to female T cells. T cell recognition of the allelic H13 peptides, therefore, lies on the cusp of foreign- versus self-reactivity. In fact, the anti-H13^a T cell recognized the foreign SVL9 and the self SIL9 peptides to equivalent levels at only a few hundred-fold difference in peptide concentration.[12] The S[V,I]L9 peptides should therefore be considered partial and reciprocal agonists that define the limit of negative selection. The ability of T cells to escape negative selection requires that the number of self peptide/MHC complexes be limited to less than a few hundred.[35,39] The abundance of the H13^a peptide is well within that range at 50 complexes/ cell (Table 4.1). However, without more data on other closely matched allelic H peptides, it difficult to generalize. Nevertheless, the *H13* locus, the gene product, and the processed peptide serve as an important paradigm for future studies.

Conclusion

The existence of H antigens has been known for over 50 years, however, the identity of these antigens at the molecular level has remained elusive due to the technical challenges of isolating T-cell stimulating antigen genes. We have developed an expression cloning strategy which we have used successfully to identify several T-cell antigens including the H13 antigen. We believe that this method has widespread application to the identification of H antigen genes and T-cell antigens in general. The identity of H13 and other *H* loci now provide the essential tools to understand their role in transplant rejection and in regulating the T-cell repertoire.

References

1. Snell GD. Methods for the study of histocompatibility genes. J Genet 1948; 49:87-108.
2. Doolittle DP, Davisson MT, Guidi JN et al. Catalog of mutant genes and polymorphic loci. In: Genetic variants and strains of the laboratory mouse. Lyon MF, Rastan S, and Brown SDM, eds. Oxford: Oxford University Press, 1996:17-854.
3. Goulmy E, Schipper R, Pool J et al. Mismatches of minor histocompatibility antigens between HLA-identical donors and recipients and the development of graft-versus-host disease after bone marrow transplantation. N Engl J Med 1996; 334:281-285.
4. Wallny H-J, Rammensee H-G. Identification of classical minor histocompatibility antigen as cell-derived peptide. Nature 1990; 343:275-278.
5. Malarkannan S, Shih P, Eden P et al. The molecular and functional characterization of a dominant minor H antigen, H60. J Immunol 1998; 160:3501-3509.
6. Shastri N, Serwold T, Gonzalez F. Presentation of endogenous peptide/MHC class I complexes is profoundly influenced by specific C-terminal flanking residues. J Immunol 1995; 155:4339-4346.
7. Del Val M, Schlicht H-J, Ruppert T et al. Efficient processing of an antigenic sequence for presentation by MHC class I molecules depends upon its neighboring residues in the protein. Cell 1991; 66:1145-1153.
8. Scott DM, Ehrmann IE, Ellis PS et al. Identification of a mouse male-specific transplantation antigen, H-Y. Nature 1995; 376:695-698.
9. Wang W, Meadows LR, den Haan JMM et al. Human H-Y: A male-specific histocompatibility antigen derived from the SMCY protein. Science 1995; 269:1588-1590.
10. Greenfield A, Scott D, Pennisi D et al. An H-YD^b epitope is encoded by a novel mouse Y chromosome gene. Nature Genet 1996; 14:474-478.
11. Bhuyan PK, Young LL, Fischer Lindhal K et al. Identification of the rat maternally transmitted minor histocompatibility antigen. J Immunol 1997;158: 3753-3760.
12. Mendoza L, Paz P, Zuberi AR et al. Minors held by majors. The H13 minor histocompatibility locus defined as a peptide/MHC class I complex. Immunity 1997; 7:461-472.
13. Simpson E, Roopenian DC. Minor histocompatibility antigens. Curr Opin Immunol 1997; 9:655-661.
14. Loveland B, Wang C-R, Yonekawa H et al. Maternally transmitted histocompatibility antigen of mice: A hydrophobic peptide of a mitochondrially encoded protein. Cell 1990; 60:971-980.
15. Morse M-C, Bleau G, Dabhi VM et al. The COI mitochondrial gene encodes a minor histocompatibility antigen presented by H2-M3. J Immunol 1996; 156:3301-3307.
16. Meadows L, Wang W, den Haan JMM et al. The HLA-A*0201-Restricted H-Y antigen contains a posttranslationally modified cysteine that significantly affects T-cell recognition. Immunity 1997; 6:273-281.
17. Markiewicz M A, Girao C, Opferman et al. Long-term T-cell memory requires the surface expression of self-peptide/major histocompatibility complex molecules. Proc Natl Acad Sci USA 1998; 95:3065-3070.
18. den Haan JMM, Meadows LM, Wang W et al. The minor histocompatibility antigen HA-1: A diallelic gene with a single amino acid polymorphism. Science 1998; 279:1054-1057.

19. Zuberi AR, Christianson G J, Mendoza LM et al. Positional cloning and molecular characterization of an immunodominant cytotoxic determinant of the mouse H3 minor histocompatibility complex. Immunity 1998; 687-698.
20. Shastri N, Needles in haystacks. Identifying specific peptide antigens for T cells. Curr Opin Immunol 1996; 8:271-277.
21. Lindahl KF, Byers DE, Dabhi VM et al. H2-M3, A full-service class Ib histocompatibility antigen. Annu Rev Immunol 1997; 15:851-879.
22. Lindahl KF. Mitochondrial inheritance in mice. Trends Genet 1985; 1:135-139.
23. Lindahl KF. Genetic variants of histocompatibility antigens from wild mice. Curr Top Microbiol Immunol 1986; 127:272-278.
24. Zuberi AR, Roopenian DC. High-resolution mapping of a minor histocompatibility antigen gene on mouse Chromosome 2. Mamm Gen 1993; 4:516-522.
25. den Haan JMM, Sherman NE, Blokland E et al. Identification of a graft versus host disease-associated human minor histocompatibility antigen. Science 1995; 268:1476-1480.
26. Skipper JCA, Hendrickson RC, Gulden PH et al. An HLA-A2 restricted tyrosinase antigen on melanoma cells results from post-translational modification and suggests a distinct antigen processing pathway for membrane proteins. J Exp Med 1996; 183:527-534.
27. Mendoza LM, Malarkannan et al. The minor H47 antigen; in preparation.
28. Van den Eynde BJ, van der Bruggen P. T cell-defined tumor antigens. Curr Opin Immunol 1997; 9:684-693.
29. Ausubel FM, Brent R, Kingston RE et al, eds. Current protocols in molecular biology. Brooklyn: Greene Publishing Associates, 1994: Vol. 1.
30. Aruffo A, Seed B, Molecular cloning of a CD28 cDNA by a high-efficiency COS cell expression system. Proc Natl Acad Sci USA 1987; 84:8573-8577.
31. Brichard V, Van Pel A, Wolfel T et al. The tyrosinase gene codes for an antigen recognized by autologous cytolytic T lymphocytes on HLA-A2 melanomas. J Exp Med 1993; 178:489-495.
32. Karttunen J, Shastri N. Measurement of ligand induced activation in single viable T cells using the lacZ reporter gene. Proc Natl Acad Sci USA 1991; 88:3972-3976.
33. Karttunen J, Sanderson S, Shastri N. Detection of rare antigen presenting cells by the lacZ T-cell activation assay suggests an expression cloning strategy for T-cell antigens. Proc Natl Acad Sci USA 1992; 89:6020-6024.
34. Malarkannan S, Afkarian M, Shastri N. A rare cryptic translation product is presented by K^b MHC class I molecule to alloreactive T-cells. J Exp Med 1995; 182:1739-1750.
35. Malarkannan S, Gonzalez F, Nguyen V et al. Alloreactive $CD8^+$ T cells can recognize unusual, rare and unique processed peptide/MHC complexes. J Immunol 1996; 157:4464-4473.
36. Malarkannan S, Serwold T, Nguyen V et al. The mouse mammary tumor virus *env* gene is the source of a $CD8^+$ T-cell-stimulating peptide presented by a major histocompatibility complex class I molecule in a murine thymoma. Proc Natl Acad Sci USA 1996; 93:13991-13996.
37. Sanderson S, Shastri N. LacZ inducible peptide/MHC specific T hybrids. Int Immunol 1994; 6:369-376.
38. Falk K, Rotzschke O, Stevanovic S et al. Allele-specific motifs revealed by sequencing of self-peptides eluted from MHC molecules. Nature 1991:290-296.
39. Kageyama S, Tsomides TJ, Sykulev Y et al. Variations in the number of peptide-MHC class I complexes required to activate cytotoxic T-cell responses. J Immunol 1995; 154:567-576.

CHAPTER 5

Mitochondrially-Encoded Minor Histocompatibility Antigens

Geoffrey W. Butcher and Lesley L. Young

The mystery of the molecular nature of minor histocompatibility (H) antigens is solved: this volume, and the meeting that begot it, record both that fact and the uses to which this new knowledge is being put. Historically the solution to the mystery is in no small part due to the study of some exotic T-cell alloantigens which proved to be controlled, in part, by the mitochondrial genome. In particular, the incisive investigations of Kirsten Fischer Lindahl and her colleagues in unpicking the unorthodox mouse Mta system (see below) led us to our first full description of a minor H antigen—which is now our description of all the rest —a self peptide bound into the peptide-binding groove of an MHC presenting molecule. That such an important immunological generalization should have emerged from the study so recherché is itself a vivid piece of evidence for the defense of the freedom of scientists to delve into dark but fascinating avenues, free of the attentions of overweening strategic science management. But this is a digression...

Today we have a clutch of minor H antigens that we know to be controlled by the mitochondrial genome (Table 5.1). Considering the relative sizes of the mitochondrial and nuclear genomes (16kb vs. 300,000kb) this number is disproportionately large. This does not, however, reflect any concentration of minor H genes in the 13 open reading frames of the mitochondrial genome. Rather, it is a consequence of practical advantages that are inherent in studying genetic factors encoded on a small extranuclear chromosomal element. Mitochondria are inherited maternally, and once maternal transmission of a trait is established there is good reason to investigate the mitochondrial genome. This genome, being so small, has been tractable to complete sequence analysis for some time[5](cf. the severer demands of gene hunting on nuclear chromosomes) and it was this circumstance that allowed the identification of the mitochondrial gene sequences controlling minor H antigens to proceed quite quickly.

The Mouse Mta System: Discovery and Description

Alloreactive mouse CTL populations of unusual, H2-unrestricted, specificity were obtained by Stockinger and Botzenhardt[6] when NZB mice were immunized against some other H2^d strains. When the genetic control of this target antigen was investigated by Fischer Lindahl and coworkers its mode of inheritance proved to be unusual—non-autosomal, non-sex-linked. Eventually maternal transmission was demonstrated and the controlling genetic element named Maternally transmitted factor (Mtf): embryo transfers and other studies established the stable, cell-autonomous expression of the antigen, reducing the

Table 5.1. Known H antigens controlled by the mitochondrial genome

Antigen Name	Presenting MHC Molecule	Mitochondrial Source Protein	Peptide Sequence	Ref.
mouse				
Mtfα	H2-M3	ND1	f-MFFIN<u>I</u>L(TL)*	1
Mtfβ	H2-M3	ND1	f-MFFIN<u>A</u>L(TL)	1
Mtfγ	H2-M3	ND1	f-MFFIN<u>V</u>L(TL)	1
Mtfδ	H2-M3	ND1	f-MFFIN<u>T</u>L(TL)	1
-	H2-M3	COI	f-MF<u>I</u>NRW	2
-	H2-D^b	?	not known	3
rat				
MTA-E	RT1-A^a	ATPase 6	ILFPSS<u>E</u>RLISNR	4
MTA-K	RT1-A^a	ATPase 6	ILFPSS<u>K</u>RLISNR	22

*f indicates an N-formylated first residue; (TL) indicates that these eighth and ninth residues are not required for minor H-antigen activity; polymorphic residues are underlined.

likelihood that an infectious agent was responsible for the antigen and implicating the mitochondrion (see ref. 7 for a review of these early studies). At the time of these studies, of course, it was a remarkable, unexplained, fact that a mitochondrial gene product could have an impact on the antigenicity of the cell surface of cells.

Genetic studies with exotic mice served to draw the Mta system back into the known realm of T-cell reactivity. It was shown by means of crosses between standard laboratory mice and Mus musculus castaneus, that the anti-Mta CTLs were MHC-restricted, not by H2-K, D or L, but by an element that was apparently non-polymorphic amongst the routine laboratory mouse strains. Castaneus mice, however, had a different form of this restriction element, and so were enlisted for the genetic analysis of this MHC-linked component (named Hmt). To cut this part of the story particularly short, the gene mapping and cloning of Hmt uncovered the fact that the restriction element for the Mta system was a non-classical class Ib MHC molecule which mapped telomeric of the classical H2 complex on mouse chromosome 17. Finally Hmt was identified[8] as the class Ib molecule now called H2-M3 which maps within a cluster of sequence-related genes in the H2-M region, marking the telomeric 'end' of the recognized MHC of rodents. Why was a strange class Ib molecule involved in this response?

But we should return to the trail of the maternally-inherited contribution to the Mta antigen. The polymorphism underlying the Mta system proved to be moderate: eventually, after extensive analysis of both laboratory and wild mice, four alleles of Mtf were described, α, β, γ, δ, which provoked distinct CTL responses. The likelihood that the mitochondrial genome was controlling Mta was increased when it was observed that mitochondrial RFLP patterns correlated with the T cell-defined Mtf alleles. With the nucleotide sequence of the Mtfα mitochondrial genome already to hand,[5] sequence comparisons with the β, γ and δ genomes were undertaken. Although a substantial collection of differences was found, in only one place within the open-reading frames did the four genomes all differ in the same codon. This was at the very 5′ end of the coding sequence for NAD dehydrogenase subunit 1 (ND1). Of course, the 1980s had seen the discovery that T cells recognize foreign antigens not as intact wholes, but in the form of short peptides. It was therefore logical to guess that the Mtf allelelic variants might be short peptides derived from mitochondrially-encoded proteins. Fischer Lindahl and colleagues proceeded to synthesize peptides corresponding to the ND1 variant sequences. Wisely they synthesized peptides with N-formyl methionine at the N-terminus: this was to reproduce the habit of mitochondria which, like bacteria, initiate their protein translation in this way. By using these peptides to sensitize appropriate CTL-target combinations they were able to show that the ND1 peptides were indeed the Mtfs in this system.[1] And so the first minor H antigen had been described. That it should turn out to be a combination of a strange presenting molecule, an unusually modified peptide (the work of Shawar et al[9] revealed the crucial importance of N-formyl modification for the biological activity of the Mtfs), and an unexpected source of antigen with unconventional inheritance does not detract from the fact that a general truth about minor H antigens had been established. These long-enigmatic antigens could now be described in the same breath as conventional T-cell antigens, the difference being that, for minors, the peptide epitopes were the products of the self's own genes and were antigenic as a result of natural polymorphism.

To round off this historical section on mouse Mta, we should describe how structural studies provided a good explanation of why the MHC class Ib molecule H2-M3 was needed in this story. Analysis of MHC class Ia-peptide complexes had emphasized the importance of conserved elements of structure that bind the amino and carboxyl termini of the peptides into the peptide binding groove.[10] The N-formylated methionine at the amino-terminus of the Mtf peptides could not be adequately accommodated in a class Ia structure. H2-M3, however, proved to have a modified groove structure which accommodates N-formyl-Met peptides excellently[11] (Fig. 5.1). Indeed, H2-M3 should be considered a specialized N-formyl-Met presentation structure and, moreover, there is now good supporting evidence for the view that it has evolved to present NfMet peptides from pathogenic bacteria (e.g., ref. 12). Rats too possess a very similar presenting molecule,[13] but, strangely, no equivalent has yet been identified in humans. One interesting consequence of the specialized structure of H2-M3 is its capacity to bind far shorter peptides than do MHC class Ia molecules, with natural ligands as short as five, six or seven residues being detected or implicated.[2,12,14]

Other Mitochondrial Minor H Antigens: Rat MTA

We now know that other genes in the mitochondrial genome can give rise to CTL-detected alloantigens. Thus an N-fMet peptide derived from the polymorphic N-terminus of cytochrome c oxidase subunit I (COI) was shown to be responsible for a minor H antigen defined by CTLs raised in (LP female x B6 male)F$_1$ mice against B6 which, again, is presented by H2-M3 (ref. 2).

Is there some necessary connection between H2-M3 and the presentation of peptide antigens derived from the mitochondrion? The answer to this question is "no." It is clear that polymorphic peptides derived from internal sequences of mitochondrial proteins (and hence not formylated) are presented in conventional fashion by classical class I molecules. This has been demonstrated for the rat maternally transmitted antigen (MTA-see below) and is presumably true for an unidentified, mitochondrially-controlled, antigen presented by H2-D^b in the mouse.[3] Thus, the special connection between H2-M3 and mitochondrial antigens relates to the availability of formyl-modified peptide sequences, and, presumably, to no other aspect of the presentation process.

No specific, mitochondrially controlled, minor H antigens have yet been reported in humans. Given the considerable variation in human mitochondria evident from RFLP and sequence analysis, it is likely that they will eventually be noticed. No evidence yet suggests that they will represent a major class of histoincompatibility, although it is quite possible that anti-human Mta responses will be generated, for instance, in bone marrow transplants between fathers and their offspring.

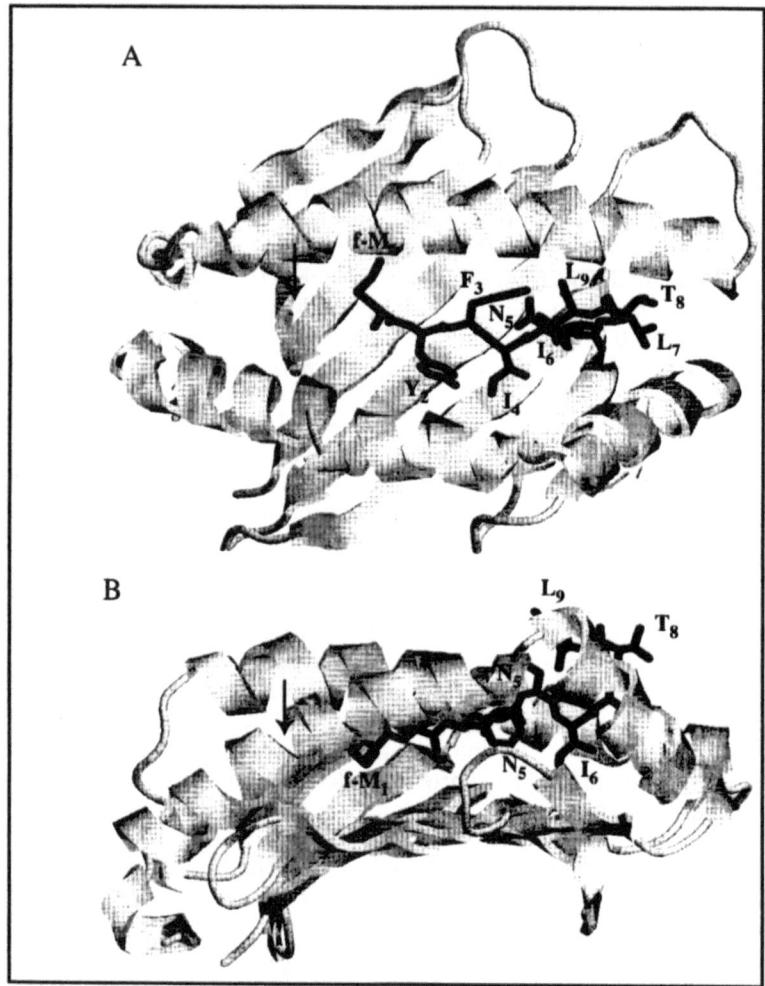

Fig. 5.1. Top (A) and side (B) views of a mitochondrial ND1 (Mtf) peptide, N-formyl-MYFINILTL bound into the peptide-binding groove of the H2-M3 class Ib molecule. This is the rat variant (P2 FcY) of Mtfα the first minor histocompatibility antigen to be fully described. Note that the peptide is positioned further to the 'right' in the groove than are peptides in class Ia molecules, which is a consequence of the specialization of this structure to accommodate N-formyl peptides. The arrows indicate the typical 'leftward' boundary of a peptide bound in the human class Ia molecule HLA-A2. The most C-terminal two residues of the antigenic peptide in this structure, which can be seen emerging above the groove in (B), are probably unnecessary for the formation of the natural minor H antigen where the peptide may be only 7 residues in length. Crystal structure coordinates were obtained from the Brookhaven National Laboratory Protein Data Bank (PDB-file name 1MHC) and modelled by means of the RASMOL Freeware software package.

Rat MTA was first uncovered in a series of experiments carried out in Darcy Wilson's laboratory, where an interest in the graft-versus-host (GVH) resistance phenomenon[15] led on to an examination of F_1 anti-parental T-cell responses in a series of rat strain combinations. Davies and colleagues analyzed in detail[16-18] a potent CTL response obtained in (LEW female x DA male)F$_1$ rats immunized against DA lymphoid cells. It became clear from both genetic segregation studies and the use of intra-MHC recombinant rat strains that this CTL response contained two separable components:

1. a CTL population specific for an antigen (termed "H") which is controlled by the class Ib region of the rat MHC (*RT1-C/E/M*) and is dependent upon homozygosity in that region for its expression.[17] Further studies of this interesting target antigen have yet to be undertaken.

2. a second CTL population whose target displayed maternal transmission.[16] Progeny of DA strain mothers were all antigen-positive. Expression of the antigen (MTA) correlated with both MHC haplotype (*RT1^{av1}*) and mitochondrial RFLP typing. Furthermore, antigen expression could be inhibited by treatment of target cells with the inhibitor of prokaryotic translation chloramphenicol, again indicative of mitochondrial involvement. Finally, regional mapping located the MHC component required in the class Ia *RT1-A* region, implicating the RT1-A^a molecule as the presenting element for the antigen.

[Later on we discuss differences in the protocols used by Davies and ourselves to generate the anti-MTA component of these CTLs.]

Identification of the Rat MTF

For our laboratory, the description of the rat MTA antigen was opportune. Our work on the rat class I modification (cim) phenomenon, which involves the antigenic alteration of an MHC class Ia molecule by an additional polymorphic MHC factor (now known to be TAP2), revolved around the expression of the *RT1-A^a* class Ia allele.[19] We were hungry for information on the peptide specificity of RT1-A^a: perhaps the rat MTA system would allow us to identify a natural peptide ligand for A^a by following the path beaten before us for mouse Mta.

In a collaboration with the Fischer Lindahl laboratory,[4] the 13 rat mitochondrial ORFs were sequenced in full from three rat strains viz. LEW (MTF$^-$), DA and BN (both MTF$^+$) and in selected regions from WF (MTF$^-$) and PVG (MTF$^+$). In this way we found four candidate coding polymorphisms located in the *ND2*, *ATPase 6*, *ND4* and *Cytb* genes, all of them due to single nucleotide changes. 17-mer peptides were synthesized around these (corresponding to the predicted sequences from DA-type mitochondria) and, by sensitizing appropriate targets in CTL experiments, it was found that the ATPase 6 candidate contained the peptide we were seeking. We quickly plumped for an 8-mer peptide within this sequence, ILFPSSER, since it corresponded well with anchor motif data that our laboratory had obtained for RT1-A^a (ref. 20). But we were wrong—this peptide did not sensitize for MTA-specific lysis. A more systematic approach led us to the conclusion that an unexpectedly long 13-mer peptide, which extended the 8-mer we had tried first by a further five amino acids in the C-terminal direction (i.e., ILFPSSERLISNR), was probably the natural ligand. This peptide, which we gave the name MTA-E, again displayed the A^a anchors but perhaps used the second arginine rather than the first as its C-terminal anchor. MTA-E is as effective a stabilizer of RT1-A^a as we have seen, both when it is expressed on the TAP-deficient cell line RMA-S and when we prepare soluble recombinant protein in *E. coli*.[4,21] The discovery that rat MTF is a 13-mer peptide provided a small but important lesson: the standard method for epitope identification, which uses panels of overlapping 9-mer peptides, would, in this case, have failed completely. The use of longer peptides (17-mers in our case), which are subject to nibbling proteolysis during the CTL assay, may be a more secure approach.

The Other Allele of Rat MTA

We have also succeeded in generating CTLs against a LEW mitochondrial antigen, again restricted by RT1-A^a (see ref. 22). A good candidate for the peptide epitope in this case is a 13-mer bearing the E to K polymorphism at the seventh position of the ATPase 6 MTA (viz. ILFPSSKRLISNR). Our published experiments show that this peptide (MTA-K) stabilizes RT1-A^a efficiently.[4]

Factors Influencing the Immunogenicity of Mitochondrial Minor H Antigens In Vivo and In Vitro

Generating specific CTL populations which recognise mitochondrial antigens has been relatively straightforward. Indeed, under optimal conditions, anti-mouse Mta CTLs have even been observed in primary in vitro cultures, e.g., in the strain combination (NZB x B10.D2)F$_1$ anti-BALB/c.[23] Notwithstanding, there are a number of factors controlling the efficacy and specificity of the response. In all the mitochondrial systems described to date in vivo priming is the standard routine for generating CTLs, as is the case for most minor H antigens. The precise conditions,viz. the site of immunization, the potential for GVH-alloreactivity by the inoculated cells and the presence or absence of additional helper epitopes carried by the immunizing cells, however, can be critical. This is particularly evident in the response to the rat MTA-E antigen.

Priming and the Immunization Route in the Anti-Rat MTA Response

Davies and coworkers clearly demonstrated that their MTA(DA-type)-specific CTL populations (described above) were raised only if:

1. the (LEW female x DA male)F$_1$ responders were primed;
2. the cells used for immunization both carried the antigen and were potentially alloresponsive to the recipient [i.e., capable of recognizing MHC alloantigens expressed by cells of the F$_1$ recipient rat and so producing a GVH reaction] and;
3. the primed lymphocyte populations were restimulated in vitro with activated lymphocytes.

In our own studies, we were keen to focus on the anti-MTA component of the Davies et al CTLs and to avoid the previously mentioned anti-"H" component. This inevitably entailed trials of various immunization strategies, some of which are illustrated in Figure 5.2. The protocol we eventually developed (Fig 5.2E) still required, like that of Davies et al, an in vivo priming step but, to our surprise, it differed in all other respects. A critical factor appeared to be the immunization route; we employed subcutaneous (s.c.) and intraperitoneal (i.p.) injections of cells in contrast to the footpad immunizations used by Davies. In our protocol, there was no requirement for potential anti-MHC GVH responsiveness by the inoculated cells (compare Figs. 5.2C and 5.2D). Thus we were able to elicit the CTL populations we wanted by immunizing (LEW female x DA male)F$_1$'s with reciprocal (DA female x LEW male)F$_1$ cells. Furthermore, in vitro restimulation simply with normal, not activated, lymph node or spleen cells was sufficient to generate good CTL populations. Importantly, the specificity of the CTL population elicited could be modified to exclude anti-H reactive cells. In addition, we observed that, when using (LEW x DA)F$_1$ anti-DA immunization, even a simple in vitro restimulation with F$_1$ cells in place of the parental DA lymphocytes was sufficient to enhance the anti-MTA component (compare Figs.5.2A and 5.2B).

Modulation of CTL specificity and strength of the response have been studied previously in the response to mouse HY. The low response mediated by CBA strain females against male can be enhanced by changing the immunization route from i.p. to footpad (N.B. apparently the opposite observation from that with the rat MTA); and by a change of inoculating cells, from splenocytes to an enriched population of adherent cells.[24,25]

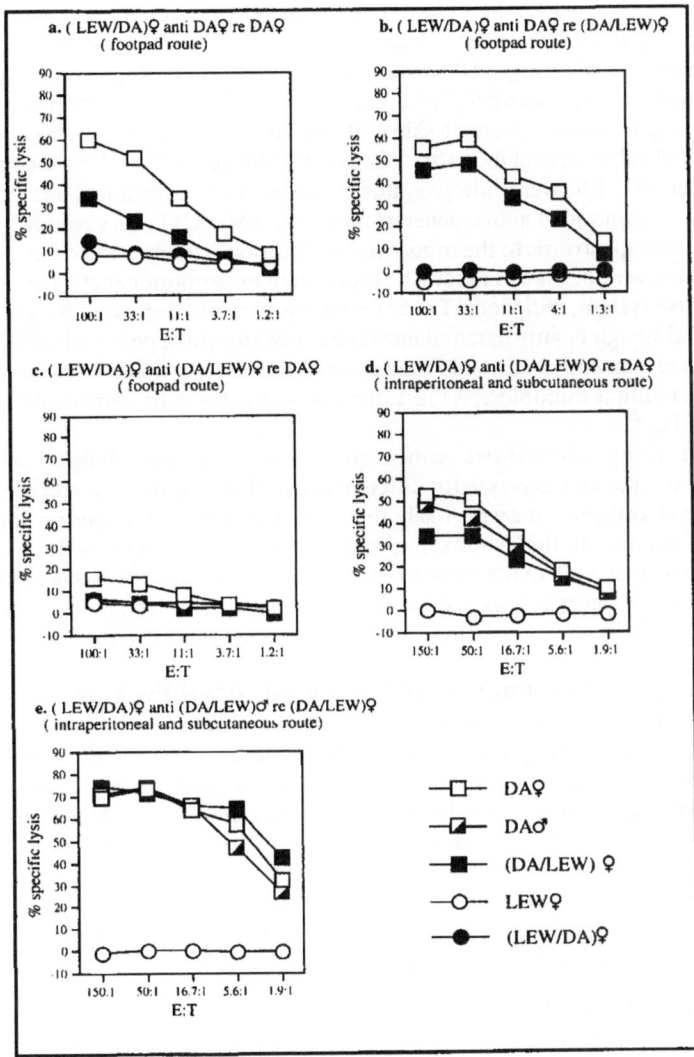

Fig. 5.2. Optimal in vivo and in vitro priming for generation of anti MTA-E CTL. This figure illustrates the different immunization and in vitro restimulation protocols used to enhance the MTA-E component of our bulk CTL populations. (LEW/DA) F₁ female rats were used in all the experiments shown. In panels A, B and C these rats were immunized with splenocytes in both rear footpads; the popliteal lymph nodes were later harvested and restimulated in vitro with splenocytes of the strain indicated. In panels D and E the animals were first immunized intraperitoneally and subcutaneously with splenocytes of the strain indicated, then the cervical and mesenteric lymph nodes and spleens removed and restimulated in vitro as indicated. Protocols are described in detail in ref. 4.

Use of Additional Minors As Helper Epitopes

We described in the previous subsection that, by immunizing (LEW female x DA male)F₁s with reciprocal (DA female x LEW male)F₁ cells, we were successful in generating anti-rat MTA(DA) CTLs. This was even true when immunizing female rats with female cells, despite the fact that in these circumstances there is only a mitochondrial disparity. We were able to improve the strength of this response, and the reciprocal response to the MTA-K allele, when the male antigen HY was used as a helper epitope on the inoculating cells, but this male 'help' was not essential (Fig. 5.2E). This is in contrast to the mouse Mta system in which priming with cells (but not a skin graft) bearing only an Mta disparity was able to lead to tolerance of subsequent skin grafts.[26] Cells that differed by several mouse minor H antigens in addition to Mta were, however, able to accelerate the rejection of later Mta-disparate grafts, and, as mentioned above, generate anti-mouse Mta CTLs very readily.[2] This suggests that the T-cell help intrinsic to the mouse mitochondrion is weak in the strain combinations used and thus the response can readily be modulated by immunogenetic (or other) means. In the mouse system, reciprocal F_1 combinations have not been used for generating anti-Mta CTLs, although F_1 anti-parental immunisations (in which potential GVH reaction can take place) are frequently successful. In general, however, the combinations used are H2 matched but differ at multiple minor loci. A Qa-1 disparity can actually be detrimental to generating anti-Mta CTLs.[23]

Given "Roopenian's Rules" concerning the need for both 'helper' and 'cytotoxic' determinants (possibly expressed at different genetic loci) for the generation of an active in vivo minor H antigen,[27] it seems likely that mouse and rat mitochondria encode 'helper' epitopes in addition to the 'cytotoxic' epitopes already defined. The evidence suggests that these may be rather weak in the mouse, but the success of F_1 anti-reciprocal F_1 immunization in the rat system suggests to us that here the help may be stronger, potentially a good (and tractable) system in which to define a minor H antigen helper determinant.

Processing and Presentation of Mitochondrial Peptides

Despite our now detailed knowledge of some mitochondrial peptides and the complexes they form with MHC class Ia and Ib molecules, their "escape" from the mitochondria and entry into the class I presentation pathway remains an intriguing enigma. It has been suggested (but not without dispute) that the mitochondrial genome persists as a separate entity because the 13 proteins it encodes are too hydrophobic to be synthesized on cytosolic ribosomes and imported into the organelle, as are the scores of other polypeptides that assemble into the complexes responsible for oxidative phosphorylation.[28] So why should a system exist for exporting protein or peptidic material from mitochondria? Is the release specific and in what cellular compartment does it occur?

Mitochondrial Location and Turnover

The polypeptides that yield the mitochondrial antigens described to date (ND1,CO1 and ATPase 6) are all subunits of large multi-membered complexes in the inner mitochondrial membrane.[28] There is a crystal structure for mammalian Complex IV, of which CO1 is part, and it may not be long before determination of the F_0 "stalk" section (containing ATPase 6), completes the ATP synthase structure—which is Complex V. But the only peptide we can definitively pinpoint at present is the CO1 peptide, which is located in the matrix of the assembled Complex IV.[29] Computer predictions for the mouse ND1 peptide and the ATPase 6 peptide suggest locations in the matrix for ND1, and partial transmembrane looping into the intermembrane space for ATPase 6 (see ref. 30) although these should be interpreted with appropriate caution.

The assembly of mitochondrially encoded polypeptide subunits with those that are nuclearly encoded and imported is a tightly regulated process which is now beginning to be elucidated. The ease of studying mitochondria in yeast where mutants can be grown on fermentable carbon sources in the absence of mitochondrial function, has led to the identification and characterization of several ATP-dependent proteolytic systems[31] which may be of significance to mitochondrial antigen processing. Their strategic location in the matrix and inner membrane permits them to perform a variety of chaperone and proteloytic functions, important in regulating the stoichiometry of the complexes and the removal of misfolded proteins of both mitochondrial and nuclear origin. The distinct proteolytic capacity of mitochondria is evident from the work of Minami and colleagues[32] who observed that different peptides can be generated from the same protein depending on the intracellular location; mitochondria or cytosol, to which it is targeted.

There is certainly rapid turnover of the proteins encoding the mitochondrial antigens since CTL detection of the ND1 and ATPase 6 peptide/MHC complexes,[4,33] as well as the unidentified, H2-D^b-presented antigen,[3] can be completely inhibited within 24 hours by chloramphenicol treatment of target cells. [Inhibition of the presentation of the CO1 peptide by chloramphenicol was not reported by Morse et al[2]]. This implies that the antigenic peptides are derived from newly synthesized proteins which may or may not have undergone assembly into their functional complexes. Furthermore, it suggests that either there are limited numbers of peptide/MHC complexes at the surface or that these are relatively unstable. The mouse system illustrates that there are at least two mitochondrial peptide/H2M3 complexes represented at the surface at a given time(in Mtf$^\alpha$/H2-M3α cells). It would be interesting to investigate their relative abundance. An alternative possibility, that turnover of intact mitochondria in autophagosomes[34] results in release of peptides long enough to provide antigenic ligands for MHC molecules, appears less likely, although it awaits direct testing. We favor the possibility that the mitochondrial minor H peptides may be generated within intact viable mitochondria and "escape" to the cytosol, perhaps through porin complexes in the outer membrane of the organelle.

Transit Through, and Out of, the Cytosol

It is now clear that the majority of class I peptides are generated in the cytosol by the proteasome and translocated by the TAP1/TAP2 heterodimer into the endoplasmic reticulum (ER) lumen for assembly into the trimeric complex presented at the cell surface.[35] The mitochondrial minor H peptides from ND1 and ATPase 6 follow at least part of this route and are transported by TAP. The presentation of the ND1 Mtfα peptide is normally (but—importantly—not always) abrogated in the absence of a functional TAP heterodimer, both in RMA-S and in TAP1 'knock-out' cells.[14,36,37]

The cytosol-to-ER transport of the peptides presented by H2-M3 constitutes an interesting case, since, as we have said, they are formylated at the N-terminus. It was previously reported[38,39] that amino-terminal modification of peptides, with methyl or acetyl groups affected severely the efficiency of translocation by TAP. It is somewhat surprising, therefore, that the formyl group of the ND1 and COI peptides does not appear to prevent their presentation. Of course, there is still a gap in our understanding of the quantitative relationship between in vitro TAP transport assays and in vivo TAP transport and peptide presentation: how poor must transport be before the assembly and presentation of a given peptide-MHC complex is effectively (i.e., functionally) prevented? A substantial reduction in TAP transport efficiency may not necessarily abrogate CTL recognition.

In the rat there are two alleles of the TAP2 polypeptide which bring about major differences in the peptide selectivity of their respective TAP heterodimers.[38,40-42] Not having genetically TAP-deficient rat cells, we were unable to determine whether or not the rat MTA-E peptide was transported by TAP. However, transport of the MTA-K form of the ATPase 6 peptide was found to be dependent on the presence of the TAP2A allele, informing us that this peptide, like the mouse ND1 peptide, was TAP-dependent (see ref. 22 and Young et al manuscript in preparation). In contrast to the data on ND1 presentation in TAP⁻ cells, presentation of the MTA-K allele is completely abolished in the context of the rat TAP2B allele, perhaps suggesting that the TAP-independent route by which some ND1 peptides appear to enter the ER may be unavailable in rats. However, the very high hydrophobicity of the mouse peptides cf. the ATPase peptides may be of significance here. TAP dependent presentation of the H2-D^b restricted mitochondrially-derived epitope, and of CO1 by M3 have not been reported upon.

In conclusion, the size of the mitochondrial peptides described to date varies widely from 6 to13 amino acids. The proteases involved in generating them are unknown. We can speculate that the job may be initiated in the mitochondria and completed in the cytosol through the involvement of the proteasome, but these details are completely unknown at present: they are under investigation for the generation of the rat ATPase 6-derived peptides.

We have certainly learned a lot from these exotic mitochondrial antigens, but the cell biology of their production remains to be elucidated.

Acknowledgments

We are grateful to James Stevens for preparing Figure 5.1. We also thank Kirsten Fischer Lindahl and Prakash Bhuyan for the happy and productive collaboration between our laboratories; Thomas Langer, for informative discussions; and the Babraham Institute small-animal facility staff for careful breeding and maintenance of our rat strains. Some of the work described was supported by the UK BBSRC via a PhD studentship to LLY and core support to GWB.

References

1. Loveland BE, Wang C-R, Yonekawa H et al. Maternally transmitted histocompatibility antigen of mice: a hydrophobic peptide of a mitochondrially encoded protein. Cell 1990; 60:971-980.
2. Morse M-C, Bleau G, Dabhi VM et al. The COI mitochondrial gene encodes a minor histocompatibility antigen presented by H2-M3. J Immunol 1996; 156:3301-3307.
3. Dabhi VM, Fischer Lindahl K. CTL respond to a mitochondrial antigen presented by H2-D^b. Immunogenetics 1996; 45:65-68.
4. Bhuyan PB, Young LL, Fischer Lindahl K et al. Identification of the rat maternally transmitted minor histocompatibility antigen. J Immunol 1997; 158:3753-3760.
5. Bibb MJ, Van Etten RA, Wright CT et al. Sequence and gene organization of mouse mitochondrial DNA. Cell 1981; 26:167-170.
6. Stockinger B and Botzenhardt U. On the T-cell hyperreactivity of NZB mice against H2-identical cells. J Exp Med 1980; 152:296-305.
7. Fischer Lindahl K, Hermel E, Loveland BE et al. Maternally transmitted antigen of mice: A model transplantation antigen. Ann Rev Immunol 1991; 9:351-372.
8. Wang C-R, Loveland BE, Fischer Lindahl K. H-2M3 encodes the MHC class I molecule presenting the maternally transmitted antigen of the mouse. Cell 1991; 66:335-345.
9. Shawar SM, Cook RG, Rodgers JR et al. Specialized functions of MHC class I molecules. I. An N-formyl peptide receptor is required for construction of the class I antigen Mta. J Exp Med 1990; 171:897-912.
10. Madden DR. The three-dimensional structure of peptide-MHC complexes. Ann Rev Immunol 1995; 13:587-622.
11. Wang C-R, Castaño AR, Peterson PA et al. Nonclassical binding of formylated peptide in crystal structure of MHC class Ib molecule. Cell 1995; 82:655-664.
12. Gulden PH, Fischer P III, Sherman NE et al. A *Listeria monocytogenes* pentapeptide is presented to cytolytic T lymphocytes by the H2-M3 MHC class Ib molecule. Immunity 1996; 5:73-79.
13. Wang C-R, Lambracht D, Wonigeit K et al. Rat RT1 orthologs of mouse *H2-M* class Ib genes. Immunogenetics 1995; 42:63-67.

14. Dabhi VM, Hovik R, Van Kaer L et al. The alloreactive T-cell response against the class Ib molecule H2-M3 is specific for high affinity peptides. J Immunol 1998; 161:5171-5178.
15. Bellgrau D, Wilson DB. Immunological studies of T-cell receptors. I. Specifically induced resistance to graft-versus-host disease in rats mediated by host T-cell immunity to alloreactive parental T cells. J Exp Med 1978; 148:103-114.
16. Davies JD, Wilson DH, Hermel E et al. Generation of T cells with lytic specificity for atypical antigens. I. A mitochondrial antigen in the rat. J Exp Med 1991; 173:823-832.
17. Davies JD, Wilson DH, Butcher GW et al. Generation of T cells with lytic specificity for atypical antigens. II. A novel antigen system in the rat dependent on homozygous expression of MHC genes of the class I-like RT1.C region. J Exp Med 1991; 173:833-839.
18. Davies JD, Wilson DH and Wilson DB. Generation of T cells with lytic specificity for atypical antigens. III. Priming F_1 animals with antigen-bearing cells also having reactivity for host alloantigens allows for potent lytic T-cell responses. J Exp Med 1991; 173:841-847.
19. Livingstone AM, Powis SJ, Diamond AG et al. A trans-acting MHC-linked gene whose alleles determine 'gain and loss' changes in the antigenic structure of a classical class I MHC molecule. J Exp Med 1989; 170:777-795.
20. Powis SJ, Young LL, Joly E et al. The rat cim effect: TAP allele dependent changes in a class I MHC anchor motif and evidence against C-terminal trimming of peptides in the ER. Immunity 1996; 4:159-165.
21. Stevens J, Wiesmüller K-H, Barker PJ et al. Efficient generation of MHC class I-peptide complexes using synthetic peptide libraries. J Biol Chem 1998; 273:2874-2884.
22. Young LL. Peptide presentation by the rat class I MHC molecule RT1.A^a. 1996; PhD Thesis (University of Cambridge, UK).
23. Smith R III, Huston DP, Rich RR. Primary cell-mediated lympholysis response to a maternally transmitted antigen. J Exp Med 1982; 156:1866-1871.
24. Müllbacher A, Brenan M. Cytotoxic T-cell response to H-Y in "non-reponder" CBA mice. Nature 1980; 285:34-36.
25. Brenan M, Müllbacher A. Analysis of the cytotoxic T-cell response to H-Y in CBA/H mice. J Immunol 1981; 127:681-685.
26. Chan T, Fischer Lindahl K. Skin graft rejection caused by the maternally transmitted antigen Mta. Transplantation 1985; 39:477-480.
27. Roopenian DC. What are minor histocompatibility loci? A new look at an old question. Immunol Today 1992; 13:7-10.
28. Darley-Usmar V, Ragan I, Smith P et al. The proteins of the mitochondrial inner membrane and their role in oxidative phosphorylation. In 'Mitochondria: DNA, Proteins and Disease' eds. Darley-Usmar V and Shapira AHV 1994, Portland Press Research Monograph.
29. Tsukihara T, Aoyama H, Yamashita E et al. The whole structure of the 13-subunit oxidized Cytochrome c Oxidase at 2.8 Å. Science 1996; 272:1136-1144.
30. Computer predictions of ND1 and ATPase6 topology were obtained using the PHD PredictProtein EMBL database. This server is described in Rost B, Sander C and Reinhard S. PHD-an automatic mail server for protein secondary structure. CABIOS 1994;10:53-60.
31. Langer T and Neupert W. Regulated protein degradation in mitochondria. Experientia 1996; 52:1069-1076.
32. Yamazaki H, Tanaka M, Nagoya M et al. Epitope selection in Major histocompatibility complex class I-mediated pathway is affected by the intracellular localization of an antigen. Eur J Immunol 1997; 27:347-353.
33. Han AC, Rodgers JR, Rich RR. An unexpectedly labile mitochondrially encoded protein is required for Mta expression. Immunogenetics 1989; 29:258-264.
34. Ashford TP, Porter KR. Cytoplasmic components in hepatic cell lysosomes. J Cell Biol 1962; 12:198-210.
35. Pamer E, Cresswell P. Mechanisms of MHC class I-restricted antigen processing. Ann Rev Immunol 1998; 16:323-358.
36. Attaya M, Jameson S, Martinez CK et al. Ham-2 corrects the class I antigen-processing defect in RMA-S cells. Nature 1992; 355:647-649.

37. Hermel E, Grigorenko E, Fischer Lindahl K. Expression of medial class I histocompatibility antigens on RMA-S mutant cells. Int Immunol 1991; 3:407-412.
38. Momburg F, Roelse J, Howard JC et al. Selectivity of MHC-encoded peptide transporters from human, mouse and rat. Nature 1994; 367:648-651.
39. Schumacher TNM, Kantesaria DV, Heemels MT et al. Peptide length and sequence specificity of the mouse TAP1/TAP2 translocator. J Exp Med 1994; 179:533-540
40. Powis SJ, Deverson EV, Coadwell WJ et al. Effect of polymorphism of an MHC-linked transporter on the peptides assembled in a class I molecule. Nature 1992; 357:211-215.
41. Momburg F, Armandola EA, Post M et al. Residues in TAP2 peptide transporters controlling substrate specificity. J Immunol 1996; 156:1756-1763
42. Deverson EV, Leong L, Seelig A et al. Functional analysis by site-directed mutagenesis of the complex polymorphism in rat transporter associated with antigen processing. J Immunol 1998; 160:2767-2779.

CHAPTER 6

The Male-Specific Minor Histocompatibility Antigen, HY

Elizabeth Simpson, Phillip Chandler and Diane Scott

HY was not the first minor histocompatibility (H) antigen to be detected in mice, since the work of several investigators during the first half of the 20th century had previously established that there were a number of independently segregating loci controlling the rejection of allogeneic skin and tumor grafts in mice (reviewed in ref.1). One of these, George Snell, was therefore well placed to interpret the findings of Eichwald and his colleagues when they reported in 1955 that in some but not all inbred strains of mice, females could reject skin grafts from males of the same strain.[2] Snell pointed out that these results were consistent with there being a minor *H* locus on the Y chromosome.[3] Subsequently, Billingham and his colleagues used HY as a model 'weak' transplantation antigen to further explore the induction of neonatal tolerance,[4,5] following the pioneering work in Peter Medawar's laboratory on neonatal tolerance to incompatibilities at H2, the major histocompatibility complex (MHC).[6-8] Exploration of the genetic control of the ability to reject syngeneic male grafts was undertaken by Bailey, another pioneer of minor H antigens: he extended the analysis of HY responder and non-responder strains and showed that a gene within the MHC was the major controlling factor.[9,10]

Investigation of minor H antigens was limited to in vivo responses until the 1970s: at the beginning of that decade attempts were made to extend the analysis by making male specific antibodies, and whilst these initially looked promising,[11] the results were difficult to repeat in other laboratories because of the low titers of the antibodies, and from studies typing individuals with abnormal Y chromosomes it appeared that the determinant identified serologically (renamed SDM) was not the same as the minor histocompatibility antigen, HY, detected by T cells.[12]

By the mid-1970s, it became possible to generate T-cell responses in vitro to HY in mixed lymphocyte reactions (MLR) using spleens of female mice that had previously rejected a syngeneic male graft or had been immunized in vivo with syngeneic male cells.[13] Male-specific cytotoxic T cells could likewise be generated in MLR using peripheral blood lymphocytes from patients undergoing host-versus-graft (HVG) or graft-versus-host (GVH) reactions following bone marrow transplantation (BMT) between HLA-identical, sex-mismatched siblings.[14] From the target cell specificity of cytotoxic cells from these MLR it was clear that these T cells were MHC-restricted in the same way as virus-specific effector cells.[15] Later work of Townsend demonstrating that the viral component of the target antigen was a small peptide[16] indicated that minor H antigens were likely also to be short peptides derived from intracellular proteins, presented by self-MHC molecules. The crystal structure of HLA-A2[17] showed how peptides were bound in a groove created by the folding of the

---

Minor Histocompatibility Antigens: From the Laboratory to the Clinic, edited by Derry Roopenian and Elizabeth Simpson. ©2000 Landes Bioscience.

alpha 1 and 2 domains of the MHC heavy chain, and knowledge of the biosynthesis of MHC class I molecules[18] provided an understanding of how peptides from endogenous proteins provided a component essential for the cell surface expression of MHC molecules. Rammensee's pioneering work (see Chapter 3) provided direct evidence for the peptidic nature of minor H antigens by eluting them from the relevant MHC molecules.[19,20]

Whilst T-cell clones specific for HY and other minor H antigens provided crucial tools for unraveling the biochemical nature of MHC peptide ligand complexes, they also provided tools for tracking down the genes encoding the peptides. One approach for HY used positional mapping for both the mouse[21-24] and human[25-27] Y chromosome genes. This allowed the identification of potential candidate genes that could then be tested by transfection.[28,29] This approach has also now been successfully used to facilitate the identification of genes encoding components of several autosomal minor H antigens (ref. 30 and summarized in Chapter 4). The initial finding of two genes encoding different HY peptide epitopes[28,29] substantiated the hypothesis put forward by Roopenian, that minor H antigens recognized in vivo tended to be genetically complex.[31] A complementary approach followed on the findings of Rammensee[19,20] and relied on the identification of the sequence of peptide eluted from MHC class I molecules and then used to sensitize target cells for T-cell recognition. When the DNA sequence of the corresponding gene is already recorded in the genomic database, this can lead to the identification of the gene[32,33] but in others this endpoint is not readily achieved.[34]

IR Gene Effects on In Vivo and In Vitro Responses to HY

It was clear from the early work on HY that all strains with the H2^b haplotype (and F1 females with an H2^b parent) are high responders, females rapidly rejecting primary syngeneic male grafts and readily generating CTL in secondary MLR.[9,13,35-37] All the non-H2^b strains are low responders, failing to reject primary male grafts, although some of them can make CTL in MLR following in vivo immunization subcutaneously in the footpad with syngeneic male spleen cells,[38] and this treatment also induces females of some strains to reject male skin grafts ("second set," i.e., following immunization).[39] The strain distribution pattern of CTL responsiveness after footpad immunization suggested the involvement of non-H2 as well as H2 linked immune response (*Ir*) genes.[38,40] The H2-linked *Ir* genes are those which encode the class I and II restriction molecules: not all alleles can present HY epitopes—for example, of H2 class I allelic variants, D^b does and K^b does not: likewise, K^d serves as a restriction molecule, but not D^d, whereas both K^k and D^k present different HY epitopes.[37] The identity and mode of action of the non-H2 *Ir* genes is not known, although the chromosomal localization of some have been identified.[40] They may alter the loading of target HY peptides into MHC restriction molecules, either quantitatively or qualitatively, or affect T-cell repertoire selection.

Interesting effects are seen when several alternative H2 restriction molecules are available, as in an F1 female responding to the F1 male. Under these circumstances, providing one of the parental strains is of the dominant high responder H2^b parent, there is a tendency for the CTL response to be limited to an allele of one of the two haplotypes, despite the fact that immunization with male cells of either parental strain can elicit a response.[41,42] For example H2^{kxb} F1 females immunized in vivo with F1 male cells make only HY/D^k and not HY/D^b-specific CTL in an MLR with male cells of F1, H2^k or H2^b origin, although such F1 females will respond with D^b restricted CTL when immunized in vivo and boosted in vitro with H2^b male cells. In contrast, H2^{dxb} F1 females immunized with F1 male cells make an HY CTL response restricted only to D^b, although capable of responding with K^d restricted HY specific CTL if immunized with H2^d male cells. This "parental preference" is a manifestation of immunodominance, also seen in situations in which multiple minor H incompatibilities are present between donor and recipient, yet the responses are limited to a few peptide determinants (see refs. 43-45 and chapters 8, 14). Understanding this mechanism is important for predicting which of a number of possible stimulatory epitopes are likely to be relevant in a clinical setting of comparable or greater genetic complexity, such as those in GVH and HVG following BMT.

HY-Specific MHC Class I and II Restricted T Cells

HY specific T cells restricted by MHC class I and class II molecules are obtained from MLR using spleens of female mice that have rejected syngeneic male grafts.[46] There has been a greater focus on class I restricted cytotoxic T cells (CTL), on the assumption that they are more obvious effectors of graft rejection. However, the ability of intra-H2 recombinant strains derived from the high responder, H2^b haplotype, to reject syngeneic male grafts correlates better with their ability to mount DTH responses to HY than to generate CTL,[47] suggesting a role of class II-restricted T cells as graft rejection effectors. It is likely that under different conditions each cell type can contribute to the graft rejection response. Both types of HY-specific clones have been essential for the chromosome mapping studies which form the essential background to cloning the genes. Having a panel of such T-cell clones from the high responder H2^b haplotype (HY/D^b, HY/A^b), and the low responder H2^k haplotype (HY/K^k, HY/D^k,HY/E^k), has been important in establishing that expression of some of these epitopes is controlled independently of the others, and that they are encoded by separate genes.[28,29]

Tolerance to HY

Reference has already been made to the in vivo work on neonatally induced tolerance to HY.[4,5] These findings were later extended to CTL responses in vitro, from which it was clear that both allogeneic as well as syngeneic male cells given to high responder, H2^b females in the neonatal period prevented not only subsequent rejection of male skin but also the ability to generate HY specific CTL. Multiparous females were likewise tolerant by both in vivo and in vitro tests.[48] These findings are consistent with exposure to reprocessed HY presented by the indirect route being able to induce tolerance, and immunization of adult females with allogeneic male cells leading to cross priming.[49] However, direct exposure of neonatal females to HY, using syngeneic but not allogeneic male cells, gives rise to a profound state of tolerance. Lymphocytes from the tolerant female when adult can be transferred to neonatal females, and they in turn are rendered tolerant to HY in vivo and in vitro.[48] Earlier studies by Weissman[50] had suggested the involvement of regulatory cells of recipient origin in such transfers. This is a characteristic of tolerance to multiple minor H antigens induced when grafts expressing these multiple minor H antigens are given under a cover of CD4 and/or CD8 antibodies by Waldmann and his colleagues.[51] These findings have been extended to the induction of linked suppression to additional minor H epitopes when they are presented on the same test graft as that expressing those minor H antigens to which tolerance was originally induced (see ref. 52 and Chapter 14). This raises the possibility of using an individual peptide HY epitope to induce tolerance to additional minor H epitopes present on a test graft. Proof of principle of this approach is provided by the results of using both the HY/D^b and the HY/K^k peptides. Female mice of the appropriate H2 haplotype pretreated with these peptides subsequently show substantially delayed rejection of either primary (H2^b) or secondary (H2^k) syngeneic male grafts, which express additional HY epitopes, in those recognized by class II-restricted, CD4 T cells (unpublished data).

Chromosomal Mapping of *HY* Genes

The existence of Y chromosomes with deletions or translocations, in individuals and cell lines, has made it possible to localize the region containing the HY genes in both mice and humans. The use of the HY specific T-cell clones restricted by different MHC alleles has been essential for these studies, which have preceded the cloning of the HY genes.

The starting point was the discovery of male mice carrying an additional copy (denoted sex-reversed or Sxr[a]) of the short arm of the Y chromosome translocated distal to the pseudoautosomal region of the Y long arm (see Fig. 6.1A and refs. 22, 53). These males produce an excess of male progeny, since one half of their X-bearing sperm also carry the Sxr[a] translocation (Fig. 6.1B). Progeny inheriting X/XSxr[a] develop as males (Fig. 6.1C) because the translocated Sxr[a] contains *Sry*, the testis determining gene.[54] They are also HY positive,[55] thus localizing both *Sry* and the *HY* gene(s) to the translocation, and hence to the short arm of the Y chromosome of normal male mice.[22] However, X/XSxr[a] males are infertile, since male meiosis cannot take place in cells with two X-chromosomes. To overcome the limitations imposed by transmission of Sxr[a] via X/YSxr[a] carrier males in each generation, such males can be crossed to females carrying another translocation, T16XH (T16X), involving chromosome 16 and the X chromosome, and the progeny of this cross includes individuals with the karyotype T16X/XSxr[a] (Fig. 6.1D and ref. 56). These mice can develop as females because preferential X inactivation of the paternal X carrying the Sxr[a] translocation spreads into the testis determining gene, *Sry*, during the crucial stage of gonadogenesis in the embryo. Such T16X/XSxr[a] females are HY positive[57] and fertile, so can be used to transmit the Sxr[a] translocation in isolation from the Y chromosome.

A further informative mutation was identified in one of these females. She and her phenotypically male progeny were also HY negative by in vitro[21] and in vivo[58] tests. This mutant, designated Sxr' and later Sxr[b], was also found to lack the functionally defined spermatogenesis gene, *Spy*, since O/XSxr[b] mice, unlike their O/XSxr[a] counterparts, failed in the early stages of spermatogenesis.[59] Several genes have been mapped to this region, including the duplicated zinc finger genes *Zfy1* and *Zfy2*,[60] *Ube1y* and its pseudogene, *Ube1yps*,[61,62] *Smcy*, a ubiquitously expressed gene identified from testis cDNA[63] and Uty, a ubiquitously expressed tetratricopeptide repeat protein homologue.[29] The Sxr[b] mutation was subsequently characterized as having undergone a fusion of the 5′ end *Zfy2* with the 3′ end of *Zfy1* with the deletion of all the intervening DNA including *Ube1y*, *Smcy* and *Uty* as well as the functionally defined *Hya* and *Spy* genes.[23] This deletion interval, ΔSxr[b], has been the focus for the identification of candidate *HY* genes.

T16X/XSxr[a] female mice provided the first evidence that there might be more than one gene responsible for HY epitope expression. Although bulk cultures from a T16X/XSxr[a] female mouse appeared to be HY-positive and express HY epitopes detected by the HY/D[b], HY/A[b], HY/D[k] and HY/K[k] specific T-cell clones, screening of individually cloned transformed bone-marrow derived cells from such a mouse showed that the expression of one epitope, HY/D[b], was subject to X-inactivation whereas that of the other three was not. This was the first evidence to suggest that expression of the HY/D[b] epitope was controlled by a separate gene.[64]

Subsequently, a deletion panel developed by imposing HY/D[b] and HY/D[k] specific CTL mediated immunoselection on an Abelson transformed B-cell clone from an O/XSxr[a], H2[kxb] F1 male provided evidence that expression of each of these four HY epitopes was separately controlled, and therefore likely to be the product of more than one gene.[24] This O/XSxr[a] deletion panel was used to identify the novel *Uty* gene[29] as a possible candidate for HY/D[b], since a *Uty*-specific cDNA probe failed to detect a male-specific band in an O/XSxr[a] deletion mutant that did not express HY/D[b].[29]

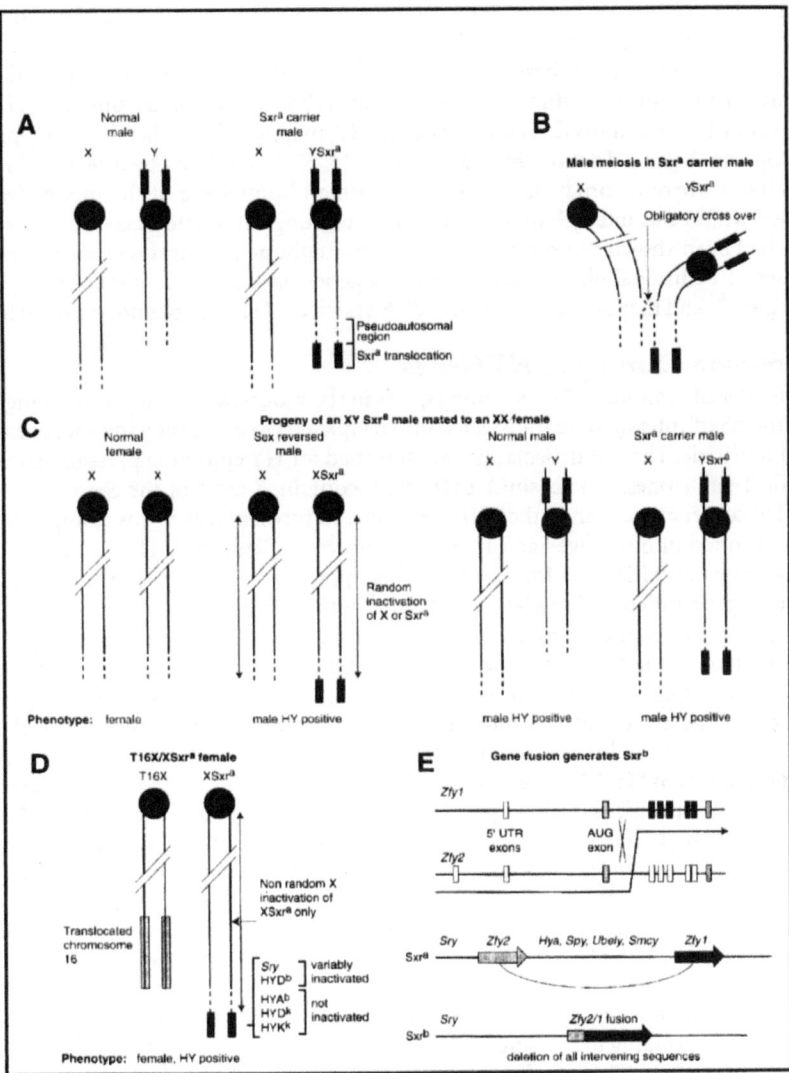

Fig. 6.1. Schematic diagram showing the genetic derivation of X/XSxr^a and T16X/XSxr^a mice; their genotype, phenotype and HY expression and the generation of Sxr^b. (A) YSxr^a was produced by a duplication of the short arm of the Y chromosome which became translocated to the pseudoautosomal region. (B) During male meiosis this duplicated region crosses over to the X chromosome to generate XSxr^a sperm. (C) The four classes of progeny from a mating between an X/YSxr^a carrier male and normal female: X/X normal female, X/XSxr^a sex reversed male, X/Y normal male and X/YSxr^a carrier male. (D) When a YSxr^a carrier male is mated with a female carrying Searle's translocation (an X-chromosome 16 translocation) the resulting T16H/XSxr^a progeny exhibit non-random X-inactivation of the paternal XSxr^a chromosome. (E) Schematic representation of the deletion event leading to the generation of Sxr^b mice. The upper part of this diagram shows details of the cross-over between the 5′ end of *Zfy2* and *Zfy1*, which takes place just 3′ of the AUG exon. This results in a *Zfy2/1* fusion product with the deletion of all the intervening genes as shown in the lower part of the figure. (Fig. 6.1E was adapted from ref. 23 by kind permission of the authors).

The localization of human HY encoding genes, *HYA*, on the long arm, Yq, of the Y chromosome, was effected by the use HLA-A2 and -B7 restricted HY specific T-cell clones to type for the presence of these epitopes on EBV transformed cell lines from patients with various deletions of the Y chromosome. Initially, *HYA* was excluded from the region of the short arm where the testis determining gene, *SRY*, mapped,[25] and then it was mapped onto the long arm, Yq, in a deletion interval (see refs. 26, 27, 65) defined by a series of naturally occuring Y chromosomal deletions.[66] The human homologue of the mouse *Smcy* gene, mapped within this interval, making this gene a strong candidate once the mouse homologue had been shown to express at least one HY epitope (see next section and ref. 25). A number of additional ubiquitously expressed genes have also recently been mapped into this region,[67] and they are candidates for HLA-restricted HY epitopes not encoded by SMCY.

Expression Cloning of *HY* Genes

To identify candidate HY encoding genes in the mouse, a series of overlapping cosmids from the ΔSxr^b interval were transfected into recipient cells expressing the appropriate MHC class I molecule. These transfectants were screened for HY epitope expression using the HY specific T-cell clones. One cosmid, cMEM 14, contained most of the *Smcy* gene, and was found to confer expression of the HY/K^k epitope. To further localize this epitope, the cosmid was subcloned into the five Sac I fragments shown in Figure 6.2. The HY/K^k epitope was expressed by the cMEM 14C fragment, which mapped to the 3′ end of the *Smcy* gene. Further subcloning identified an EcoRI-EcoRV fragment that clearly expressed the HY/K^k epitope. Despite the fact that the restriction enzyme Pst1 did not cleave near the cognate peptide (see below), none of the four subcloned Pst1 fragments was able to express HY/K^k. Previous work, examining the expression of tumour transplantation antigens from subgenic DNA fragments,[68] indicated that both intronic sequences preceding the exon expressing the epitope and in-frame ATG codons within the exon itself were required for epitope expression. The unexpected loss of HY/K^k expression following Pst1 digestion could therefore be explained by the loss of intronic sequences preceding this exon, or to the loss of 4 of the 5 in-frame ATGs, which might be required for translational initiation of this fragment. Comparison of the amino acid sequences encoded by the exons in this fragment with the equivalent region of Smcx, the X-chromosome homologue of the *Smcy* gene,[69] indicated that there were six peptides with a potential H2-K^k binding motif[70] that differed between *Smcy* and *Smcx*. One of these peptides, TENSGKDI, maximally stimulated the HY/K^k specific T-cell clone when added to female cells at nanomolar concentrations. The amino acid sequences of *Smcy* and *Smcx* in the region where the H2-K^k epitope is expressed are very divergent and a synthetic peptide corresponding to the Smcx peptide was unable to stimulate the T-cell clone, even at high concentrations. These findings thus demonstrated the molecular basis for the antigenic difference that causes this anti-HY response. The peptide encoded by *Smcy* is able to bind to the K^k class I molecule and is sufficiently different from the Smcx-encoded peptide to elicit a male-specific T-cell response that contributes towards graft rejection.

The same approach has been used to define other male-specific peptides within *Smcy*. Transfection of cMEM 14 into recipient cell lines expressing the D^k molecule were found to confer expression of the HY/D^k epitope and this was further localized to the cMEM 14B fragment (Fig. 6.2) that mapped immediately 5′ of cMEM 14C. Further subcloning using a series of PCR-amplified cDNA fragments localized the HY/D^k epitope to an approximately 300bp fragment. Although the D^k binding motif has not been fully characterized, comparison of the sequences of *Smcy* and *Smcx* within this fragment identified only six amino acid differences and these could be incorporated into four synthetic 8-12mer peptides. Only one of these stimulated significantly, although still only at much higher (micromolar in

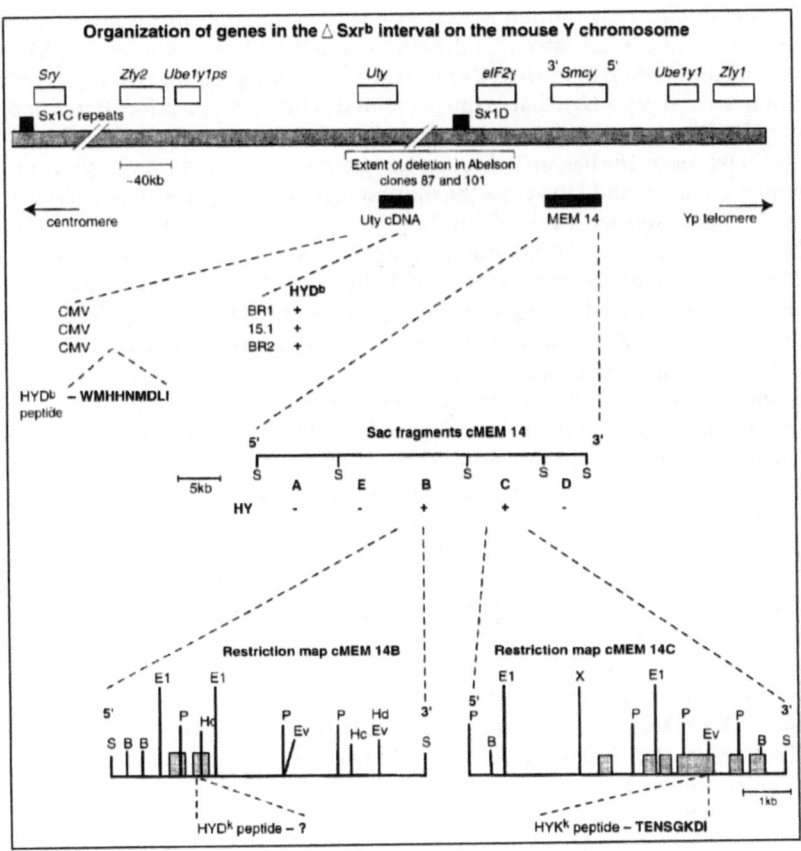

Fig. 6.2. The organization in the ΔSxr^b interval of the short arm of the mouse Y chromosome. The extent of the region is indicated by the thick grey line and the slashed lines represent breaks of unknown length. Arrows indicate the direction of the centromere and telomere. The positions of indicated genes mapping close to *Smcy* are indicated by open boxes. The positions of the Y-specific anonymous markers, Sx1C and Sx1D, are shown as the filled box above the line and those of *Uty* cDNA and cosmid cMEM 14 as the filled boxes beneath the line. The extent of the deletion in Abelson clones 87 and 101 indicated. Identification of HY epitopes within *Uty* cDNA and cMEM 14: three overlapping fragments from *Uty* cDNA, driven by the CMV promotor, all expressed the HY/D^b epitope indicated by "+." The cognate HY/D^b peptide from the N terminal region is WMHHNMDLI. Subcloning of cMEM14: the cosmid was subcloned into five SacI (S) fragments (cMEM14 A-E) which included all but 1.5 kb of the cosmid DNA. The order of these subfragments was established by hybridization with cDNAs of different lengths. The HY/D^k typing of these fragments is indicated by "+" positive and "-" negative. Restriction site and exon mapping of cMEM14B and C: BamHI (B), EcoRI (EI), EcoRV (EV), PstI (P) and XhoI (X). Exons were positioned by sequence comparison with the human X homologue, XE169 cDNA[73] and a *Smcy* cDNA. An unknown HY/D^k peptide maps within a 300 bp fragment of MEM 14B. The peptide TENSGKDI, comprising the HY K^k epitope, maps within the P and Ev restriction sites in the fourth exon of MEM 14C.

comparison to nanomolar) concentrations (unpublished observations) and work is currently underway using a series of overlapping peptides to further define this peptide epitope.

Mapping studies using immunoselected cloned Abelson transformed B-cell lines from an O/XSxr^a male mouse indicated that the HY/D^b and the HY/A^b epitopes were likely to be located within the deletion found in cell lines 87 and 101 (see Fig. 6.2 and ref. 24). Overlapping cDNAs from a novel gene, *Uty*, that mapped to this deletion were therefore tested and found to confer HY/D^b expression when transfected into recipients expressing the D^b class I molecule.[29] Like *Smcy*, *Uty* has an X chromosome homologue and, as previously, comparison of the sequences of *Uty* and *Utx* across the shortest region that expressed the HY/D^b epitope, identified six peptides with the D^b binding motif. A single peptide, WMHHNMDLI, stimulated the HY/D^b-specific clone at nanomolar concentrations. The equivalent peptide encoded by *Utx* differed by only 3 amino acids, but two of these (N at position 5 and I at position 9) are the so-called anchor residues that are required for binding to D^b.[70] Indeed it was found that both the X-encoded peptide and Y-encoded peptides that had been modified to express the X-encoded amino acids at positions 5 and 9 or 5 alone failed to bind to D^b. The third amino acid residue that differed, M at position 6, was shown to be critical for T-cell recognition.[29] Since earlier studies had indicated that HY/D^b was likely to be encoded by a gene that was separate from genes encoding the HY/K^k, HY/D^k and HY/A^b epitopes,[64] and since the overlapping cDNAs did not cover the whole Uty gene, they were not screened for HY/A^b expression. Another cDNA encoding a translational initiation factor, *eIF28γ* also mapped to this region and identified one of the breakpoints in cell lines 87 and 101 (see Fig. 6.2, ref. 71). This gene is also ubiquitously expressed and has an X-chromosome homologue; however, to date no convincing evidence that it expresses the HY/A^b epitope has been obtained.

Expression cloning in the mouse has therefore led to the identification of two genes *Smcy* and *Uty* that between them express three of the MHC class I restricted HY epitopes. However, the gene or genes responsible for expression of the MHC class II epitopes, HY/A^b, HY/A^k and HY/E^k remain to be identified. Although the HY/A^b epitope maps to the deletion interval defined by the Abelson cell lines 87 and 101, the only other candidate gene, *eIF28γ* that has so far been mapped to this region failed to express this epitope.

The human homologue of *Smcy*, *SMCY*, has also been shown to express the human HY epitopes recognized in association with the HLA-B7 and HLA-A2 (A*0201) class I molecules (refs. 32, 33; Chapter 8). This has been demonstrated using a different approach, that of peptide elution from purified MHC class I molecules followed by microcapillary liquid chromatography and electrospray ionisation mass spectrometry. The peptides that were able to mimic the HY/B7 and HY/A2 epitopes were identified originally from the mass spectrometry data. The sequences of the cognate peptides were subsequently confirmed by comparison with the known *SMCY* gene and by direct testing of synthetic peptides corresponding to these sequences for HY epitope expression.

Conclusion

HY has been a useful model minor transplantation antigen to explore this category of 'weak' antigens which elicit essentially only T-cell responses. These might be considered to include a substantial group of tumor antigens (e.g., MAGE, ref. 72; Chapter 17), and autoantigens. Since genes encoding HY epitopes are genetically isolated on the Y chromosome, they were in some ways easier to identify compared with autosomally encoded minor H antigens. It is of interest that now, when the first autosomal minor H genes are being identified, how many of them, like those encoding HY, are novel and of unknown function. *Smcy*, encoding HY/K^k [28] and HY/D^k, and *Uty*, encoding HY/D^b [29] are examples of these. The *Ir* gene control of HY by MHC class I and II molecules showed the importance of the restriction molecules in determining in vivo and in vitro responses, although other, non-MHC *Ir* genes have a modifying effect,[38,40] by mechanism(s) not yet understood. The MHC-restricting molecules play a central role in the control of immunodominance, i.e., hierarchy of responsiveness in the presence of multiple minor H disparities. It is important to understand the molecular mechanisms of this powerful biological effect, to better predict the strength of GVH and HVG in the context of particular minor H mismatches in individuals in possession of particular HLA alleles. The potential use of defined minor H peptides to modify the in vivo response is likely to be explored empirically in humans, but in mice we have the opportunity to use HY peptides to model the effect on graded additional histoincompatibilities, and attempt to learn the rules for obtaining clinically useful effects.

References

1. Snell GD, Stimpfling JH. The genetics of tissue transplantation. In: Green EL, ed. Biology of the laboratory mouse. New York: McGraw Hill, 1996:457-491.
2. Eichwald EJ, Silmser CR. Untitled. Transplant Bull 1955; 2:148-149.
3. Snell GD. A comment on Eichwald and Silmser's communication. Transplant Bull 1956; 3:29-31.
4. Billingham RE, Silvers WK. Studies on the tolerance of the Y-chromosome antigen in mice. J Immunol 1960; 85:14-26.
5. Billingham RE, Silvers WK, Wilson DB. A second study on the HY transplantation antigen in mice. Proc R Soc Lond B Biol Sci 1965; 163:61-89.
6. Billingham RE, Brent L, Medawar PB. Actively acquired tolerance of foreign cells. Nature 1953; 172:603-606.
7. Billingham RE, Brent L, Medawar PB. Quantitative studies on tissue transplantation immunity II. The origin, strength and duration of actively and adoptively acquired immunity. Proc R Soc Lond B Biol Sci 1954; 143:58-80.
8. Billingham RE, Brent L, Medawar PB. Quantitative studies on tissue transplantation immunity III. Actively acquired tolerance. Phil Trans Roy Soc B 1956; 239:357-414.
9. Bailey DW. Allelic forms of a gene controlling the female immune response to the male antigen in mice. Transplantation 1971; 11:426-428.
10. Bailey DW, Hoste J. A gene governing the female immune response to the male antigen in mice. Transplantation 1971; 11:404-407.
11. Goldberg EH, Boyse EA, Scheid M et al. Production of HY antibody by female mice that fail to reject male skin. Nature 1972; 238:55-57.
12. Simpson E. HY and sex reversal: Minireview. Cell 1986; 44:813-814.
13. Gordon RD, Simpson E, Samelson LE. In vitro cell-mediated immune responses to the male-specific (HY) antigen in mice. J Exp Med 1975; 142:1108-1120.
14. Goulmy E, Termijtelen A, Bradley BA et al. Y-antigen killing by T cells of women is restricted by HLA. Nature 1977; 266:544-545.
15. Zinkernagel RM, Doherty PC. Restriction of in vitro mediated cytotoxicity in lymphocytic choriomeningitis within a syngeneic or semi-allogeneic system. Nature 1974; 248:701-702.
16. Townsend ARM, Rothbard J, Gotch FM et al. The epitopes of influenza nucleoprotein recognised by cytotoxic T lymphocytes can be defined by short synthetic peptides. Cell 1986; 44:959-968.
17. Bjorkman PJ, Saper MA, Samraoui B et al. Structure of the human class I histocompatibility antigen, HLA-A2. Nature 1987; 329:506-512.
18. Townsend A, Elliott T, Cerundolo V et al. Assembly of MHC class I molecules analyzed in vitro. Cell 1990; 62:285-295.
19. Rötzschke O, Falk K, Wallny H-J et al. Characterization of naturally occurring minor histocompatibility peptides including H-4 and HY. Science 1990; 249:283-287.
20. Wallny H-J, Rammensee HG. Identification of classical minor histocompatibility antigen as cell-derived peptide. Nature 1990; 343:275-278.
21. McLaren A, Simpson E, Tomonari K et al. Male sexual differentiation in mice lacking HY antigen. Nature 1984; 312:552-555.

22. McLaren A, Simpson E, Epplen JT et al. Location of the genes controlling HY-antigen expression and testis determination on the mouse Y chromosome. Proc Natl Acad Sci USA 1988; 85:6442-6445.
23. Simpson EM, Page DC. An interstitial deletion in mouse Y-chromosomal DNA created a transcribed Zfy fusion gene. Genomics 1991; 11:601-608.
24. King TR, Christianson GJ, Mitchell MJ et al. Deletion mapping using immunoselection for HY further resolves the Sxr region of the mouse Y chromosome and reveals complexity at the Hya locus. Genomics 1994; 24:159-168.
25. Simpson E, Chandler P, Goulmy E et al. Separation of the genetic loci for the HY antigen and testis determination on the human Y chromosome. Nature 1987; 326:876-878.
26. O'Reilly AJ, Affara NA, Simpso E et al. A molecular deletion map of the Y-chromosome long arm defining X and autosomal homologous regions and the localisation of the HYA locus to the proximal region of the Yq euchromatin. Hum Mol Genet 1992; 1:379-385.
27. Cantrell MA, Bogan JS, Simpson E et al. Deletion mapping of HY antigen to the long arm of the human Y chromosome. Genomics 1992; 13:1255-1260.
28. Scott DM, Ehrmann IE, Ellis PS et al. Identification of a mouse male-specific transplantation antigen, HY. Nature 1995; 376:695-698.
29. Greenfield A, Scott D, Pennisi D et al. An HYDb epitope is encoded by a novel mouse Y-chromosome gene. Nat Genet 1996; 14:474-478.
30. Mendoza LM, Paz P, Zuberi A et al. Minors held by majors. The H13 minor histocompatibility locus defined as a peptide/MHC class I complex. Immunity 1997; 7:461-472.
31. Roopenian DC. What are minor histocompatibility loci? A new look to an old question. Immunol Today 1992; 13:7-10.
32. Meadows L, Wang W, den Haan JMM et al. The HLA-A*0201-restricted HY antigen contains a posttranslationally modified cysteine that significantly affects T-cell recognition. Immunity 1997; 6:273-281.
33. Wang W, Meadows LR, den Haan JMM et al. Human HY: A male-specific histocompatibility antigen derived from the SMCY protein. Science 1995; 269:1588-1590.
34. den Haan JMM, Sherman NE, Blokland E et al. Identification of a graft versus host disease-associated human minor histocompatibility antigen. Science 1995; 268:1476-1480.
35. Simpson E, Gordon R. Responsiveness to HY antigen: *Ir* gene complementation and target cell specificity. Immunol Rev 1977; 35:59-75.
36. Gordon R, Simpson E. Immune response gene control of cytotoxic T-cell responses to HY. Transplant Proc 1977; 9:885-888.
37. Simpson E. The role of HY as a minor transplantation antigen. Immunol Today 1982; 3:97-106.
38. Fierz W, Brenan M, Mullbacher A et al. Non-H2 and H2 linked immune response genes control the cytotoxic T-cell response to *HY*. Immunogenetics 1982; 15:261-270.
39. Simpson E, Chandler P, Liew FY et al. Induction and effector function of T cells. In: Moller G, Moller E, eds. Genetics of the immune response. New York:Plenum Press, 1983:121-128.
40. Fierz W, Farmer GA, Sheena JH et al. Genetic analysis of the non-H-2 linked *Ir* genes controlling the cytotoxic T-cell response to HY in H2d mice. Immunogenetics 1982; 16:593-601.
41. Gordon R, Samelson L, Simpson E. Selective response to HY antigen by F1 female mice sensitised to F1 male cells. J Exp Med 1977; 146:606-610.
42. Brenan M, Simpson E, Mullbacher A. Analysis of haplotype preference in the cytotoxic T-cell response to *HY*. Immunogenetics 1981; 13:133-146.
43. Wettstein PJ. Immunodominance in the T-cell response to multiple non-HY histocompatibility antigens. II. Observation of a hierarchy among dominant antigens. Immunogenetics 1986; 24:24-31.
44. Yin L, Poirier G, Neth O et al. Few peptides dominate CTL responses to single and multiple minor histocompatibility antigens. Int Immunol 1993; 5:1003-1009.

45. Wolpert E, Franksson L, Kärre K. Dominant and cryptic antigens in the MHC class I restricted T-cell response across a complex minor histocompatibility barrier: Analysis and mapping by elution of cellular peptides. Int Immunol 1995; 7:919-928.
46. Tomonari K. Antigen and MHC restriction specificity of two types of cloned male-specific T-cell lines. J Immunol 1983; 131:1641-1645.
47. Liew FY, Simpson E. Delayed type hypersensitivity responses to HY: Characterization and mapping of *Ir* genes. Immunogenetics 1980; 11:255-266.
48. Simpson E, Benjamin D, Chandler P. Non-responsiveness to HY: Tolerance in H2b mice. Transplant Proc 1981; XIII 4.
49. Gordon RD, Mathieson BJ, Samelson LE et al. The effect of allogeneic presensitization on HY-graft survival and in vitro cell mediated responses to HY antigen. J Exp Med 1976; 144:810-820.
50. Weissman IL. Transfer of tolerance. Transplantation 1973; 15:265-269.
51. Chen ZK, Cobbold SP, Waldmann H et al. Amplification of natural regulatory immune mechanisms for transplantation tolerance. Transplantation 1996; 62:1200-1206.
52. Davies JD, Leong LY, Mellor A et al. T-cell suppression in transplantation tolerance through linked recognition. J Immunol 1996; 156:3602-3607.
53. Cattanach BM, Pollard CE, Hawkes SG. Sex reversed mice: XX and XO males. Cytogenetics 1971; 10:318-337.
54. Gubbay J, Collignon J, Koopman P et al. A gene mapping to the sex-determining region of the mouse Y chromosome is a member of a novel family of embryonically expressed genes. Nature 1990; 346:245-250.
55. Simpson E, Edwards P, Wachtel SS et al. HY antigen in Sxr mice detected by H2 restricted cytotoxic cells. Immunogenetics 1981; 13:355-358.
56. McLaren A, Monk M. Fertile females produced by inactivation of an X chromosome in 'sex-reversed' mice. Nature 1982; 300:446-448.
57. Simpson E, McLaren A, Chandler P et al. Expression of HY antigen by female mice carrying Sxr. Transplantation 1984; 37:17-21.
58. McLaren A, Hunt R, Simpson E. Absence of any male specific antigen recognized by T lymphocytes in X/XSxr' male mice. Immunology 1988; 63:447-449.
59. Burgoyne PS, Levy ER, McLaren A. Spermatogenic failure in male mice lacking HY antigen. Nature 1986; 320:170-172.
60. Mardon G, Page DC. The sex determining region of the mouse Y chromosome encodes a protein with a highly acidic domain and 13 zinc fingers. Cell 1989; 56:765-770.
61. Kay GF, Ashworth A, Penny GD et al. A candidate spermatogenesis gene on the mouse Y chromosome is homologous to ubiquitin-activating enzyme E1. Nature 1991; 354:486-489.
62. Mitchell MJ, Woods DR, Tucker PK et al. Homology of a candidate spermatogenic gene from the mouse Y chromosome to the ubiquitin-activating enzyme E1. Nature 1991; 354:483-486.
63. Agulnik AI, Mitchell MJ, Lerner JL et al. A mouse Y-chromosome gene encoded by a region essential for spermatogenesis and expression of male specific minor histocompatibility antigens. Hum Mol Genet 1994; 3:873-878.
64. Scott D, McLaren A, Dyson PJ et al. Variable spread of X inactivation affecting the expression of different epitopes of the *Hya* gene product in mouse B-cell clones. Immunogenetics 1991; 33:54-61.
65. Chandler, P. Ph.D. Thesis, 1994.
66. Vollrath D, Foote S, Hilton A et al. The human Y chromosome: A 43 interval map based on naturally occurring deletions. Science 1992; 258:52-59.
67. Lahn BT, Page DC. Functional coherence of the human Y chromosome. Science 1997; 278:675-680.
68. Chomez P, De Plaen ED, Van Pel A et al. Efficient expression of tum⁻ antigen P91A by transfected subgenic fragments. Immunogenetics 1992; 35:241-252.
69. Agulnik AI, Mitchell MJ, Mattei MG et al. A novel *X* gene with a widely transcribed Y-linked homologue escapes X-inactivation in mouse and human. Hum Mol Genet 1994; 3:879-884.

70. Rammensee HG, Falk K, Rötzschke O. Peptides naturally presented by MHC class I molecules. Annu Rev Immunol 1993; 11:213-244.
71. Ehrmann E, Ellis P, Mazeyrat S et al. The structural genes for the translation initiation factor *eIF-2γ* are on the X and Y chromosomes in the mouse but only the X chromosome in the human. Hum Mol Genet 1998; 7:1725-1737.
72. de Plaen E, Arden K, Traversari C et al. Structure, chromosomal localization and expression of 12 genes of the MAGE family. Immunogeneticts 1994; 40:360-369.
73. Wu J, Ellison J, Salido E et al. Isolation and characterization of XE169, a human gene that escapes X-inactivation. Hum Mol Gen 1994; 3:153-160.

CHAPTER 7

Mechanisms and Implications of Immunodominance

Claude Perreault, Stéphane Pion and Denis C. Roy

Definition of Immunodominance

Minor Histocompatibility (H) Antigens: Number and Polymorphism

Minor H antigens have two fundamental characteristics:
1. they are MHC-associated self peptides derived from the partial proteolysis of endogenous proteins, and
2. they are polymorphic and immunogenic for T cells.[1-7] Because of their polymorphism and immunogenicity, minor H antigens can trigger graft rejection or graft-versus-host disease (GVHD), and have therefore attracted considerable clinical interest.[1,2,7,8]

In order to assess the magnitude of the problem posed by minor H antigens in transplantation, the first questions that were addressed concerned the number of minor H loci and their level of polymorphism. As outlined in several chapters in this volume, minor H antigens generally originate from rare mutations affecting evolutionarily conserved protein sequences, and their polymorphism is low, most of them being encoded by biallelic loci.[9-13]

The question of the total number of minor H antigens is a complex issue and is still a matter of dispute. Theoretically, as a mutation in any gene could give rise to a new minor H antigen, the total number of minor H antigens harbored by an individual could be enormous.[5] However, not every polymorphic sequence is expected to be immunogenic. In order to be immunogenic, polymorphic self-peptides must be adequately presented and recognized. This means that such a peptide must:
1. be liberated from its precursor and not be destroyed,
2. compete successfully with other peptides for binding to MHC molecules, and
3. the resulting peptide/MHC complex must be recognized by a T-cell receptor (TCR).[14-16]

Furthermore, simultaneous presentation of both class I- and class II-associated epitopes on the same antigen presenting cell (APC) greatly influences the amplitude and consequences of T-cell activation.[17,18] Estimates based on the frequency of minor H gene mutations and on the segregation of independent minor H genes in congenic mice suggest that two unrelated strains of mice, such as C57BL/6 and BALB.B, differ by at least 45 minor H loci, and that the total number of minor H genes could be in the range of 430-720.[1,19] It was therefore a great surprise when CTL responses raised between different strains of mice were found to be targeted to an extremely limited number of minor H antigens, commonly 2 or 3.[20-27] This phenomenon termed "immunodominance" signifies that some minor H antigens, said to be "immunodominant," inhibit T-cell responses to nondominant minor H antigens.

Immunodominance Limits the Repertoire of T-Cell Responses

Immunodominance has also been found in immune responses to viruses, tumor antigens and autoantigens.[14,28-32] Thus, it is not peculiar to minor H antigens, but rather represents the pattern of response usually shown by T cells exposed to numerous immunogenic epitopes. A priori, it appears surprising that the immune system elects to focus its responses on such a small number of epitopes, because this could increase the risk that pathogens might evade immune surveillance (but see 'The Mechanisms of Immunodominance' in a subsequent section). However, in the context of transplantation, it may be possible to take advantage of this phenomenon. Indeed, it seems logical to assume that biologically relevant anti-minor H antigen T-cell responses should be more easy to prevent or modulate if they are limited to only a small number of epitopes.

Immunodominance In Vitro and In Vivo

CTL Responses

Cytotoxicity assays have been the most widely used read out system to dissect anti-MiHAs responses because they are convenient, highly sensitive, and also because cytotoxicity is an important effector mechanism in vivo. Thus, it is not surprising that the analysis of in vitro cytotoxic T-lymphocyte (CTL) responses provided the first evidence that immunodominance characterizes anti-minor H antigen responses. By definition, all minor H antigens, dominant or not, can elicit T-cell responses since rejection of non-MHC differences is mediated solely by T cells. However, dominant minor H antigens trigger CTL responses when presented alone or with other minor H antigens, whereas nondominant minor H antigens can only elicit CTL responses in the absence of dominant minors.[20-27] Thus, while C57BL/6 and BALB.B mice differ at more than 45 independent minor *H* loci, very few minor H antigens are targeted by C57BL/6-derived anti-BALB.B CTLs.

The Repertoire of Anti-Minor H Antigen T-Cell Responses In Vivo

To understand how T cells respond to allogeneic minor H antigens following BMT, their fate, as well as their T-cell receptor (TCR) usage and functional activity, were studied in LP mouse strain recipients of C57BL/6 marrow grafts. With this strain combination, recipients developed a severe GVHD. During the first 15 days post-transplant, a dramatic expansion (approximately equal to 10^5-fold) of donor CD8$^+$ T cells reactive to host minor H antigens took place.[33] Flow-cytometric analysis and nucleotide sequencing of the variable-diversity-joining (V-D-J) junctional region showed that this expansion primarily involved one or a few clones using Vβ5.1 or Vβ8.1 TCR elements. Expanded T-cell populations displayed in vitro MHC-restricted cytotoxicity toward a very low number of dominant minor H antigens, which were found in only two peaks following fractionation of LP minor H antigens by HPLC.[33] These observations provide strong evidence that anti-host CD8$^+$ T-cell responses are oligoclonal in nature and selectively targeted to a few minor H antigens. In addition, analysis of the TCR Vβ repertoires of CD4$^+$ thoracic duct lymphocytes collected during the initial stages of GVHD in the B6→BALB.B and B6→CXBE strain combinations suggested that, as for CD8$^+$ cells, in vivo expansion of CD4$^+$ anti-host T cells may only involve a few Vβ subsets.[34] These findings can probably be extended to humans because molecular analysis of the TCR repertoire in BMT recipients detected expansion of oligoclonal T-cell populations selectively at the site of GVHD lesions.[35]

Biologic Consequences of T-Cell Responses Toward Dominant and Nondominant Minor H Antigens

Although dominant and nondominant minor H antigens can elicit similar in vitro CTL responses when presented separately, T cells may react differently in vivo. In mice, repeated attempts to elicit GVHD by injection of either naïve or presensitized T cells specific for nondominant minor H antigens, such as H-Y or H-3, have yielded negative results.[36,37] Similarly, H-Y incompatibility was found not to influence the risk of GVHD in human.[7] In contrast, in at least two cases, T-cell responses towards a single dominant minor H antigen were sufficient to cause GVHD. In mouse, C3H.SW-derived T cells specific for the dominant D^b-associated peptide AAPDNRETF (B6^{dom}1) caused thymic hypoplasia and skin lesions in C57BL/6 recipients.[38] Furthermore, albeit immunodominance is harder to study in humans, a mismatch of the (probably dominant) HA-1 minor H antigen was found to be associated with an increased risk of GVHD.[39]

Unanswered Questions

Together, these results provide strong evidence that immunodominance limits the repertoire of anti-minor H antigen-specific T-cell responses in vivo. Moreover, they suggest that although it may not be evident when looking at in vitro CTL responses, certain minor H antigens seem to be much more effective than others at generating biologically significant in vivo responses. However, the majority of studies on immunodominance have measured the repertoire of CD8$^+$ T-cell responses. Based on the limited available data, it cannot be determined whether the similar immunodominance rules do apply to CD4$^+$ T cells.[34,40] The analysis of in vivo anti-minor H antigens T-cell responses is also hampered by our limited knowledge on the tissue distribution of minor H antigens. Whereas some minor H antigens are probably ubiquitous, the tissue expression of others is more restricted.[13,41] Importantly, molecular definition of class I- and class II-associated minor H antigens, and assessment of their tissue distribution should be instrumental towards explaining an apparent paradox: in the C57BL/6 vs. BALB.B strain combination, immunodominant minor H antigens recognized in in vitro CTL assays segregate independently (in CXB recombinant strains) from the immunodominant targets of GVHD.[42] A likely explanation is that these assays explore different universes or peptides (e.g., MHC class I and class II restricted) and potentially different effector mechanisms. Standard anti-minor H antigens CTL assays most readily detect perforin-mediated cytotoxicity towards hematopoietically-expressed minor H antigens, whereas GVHD could be a consequence of FasL, TNF and perforin-mediated CD4$^+$- and CD8$^+$-mediated effector mechanisms directed against antigens expressed on hematopoietic and epithelial cells.[8]

Mechanisms of Immunodominance

Dominant Minor H Antigens Suppress CTL Responses Towards Nondominant Minor H Antigens Presented on the Same APC

Comparison of B6^{dom}1 (AAPDNRETF) and the *Uty*-encoded HY (WMHHNMDLI) peptide epitopes in H2^b mice proved to be a particularly informative model to decipher the mechanisms of dominance because:

1. both minor H antigens are presented by the same class I molecule (D^b),
2. their biochemical sequence is known,[38,43] and
3. they lie at opposite ends on the scale of dominance. Indeed, H-Y is always nondominant when presented with one or several autosomal minor H antigens, whereas in many strain combinations B6^{dom}1 is dominant when presented with numerous other minor H antigens (refs. 26, 44; Pion S, unpublished observations).

Thus, when C3H.SW female mice are primed with C57BL/6 male cells they generate anti-B6^{dom}1 CTLs but do not respond to HY even though:

1. C3H.SW female mice can respond to HY when presented alone (on C3H.SW male cells), and
2. B6^{dom}1 did not suppress anti-HY responses by acting as a TCR antagonist for anti-H-Y CTLs because B6^{dom}1 and HY are not crossreactive at the TCR level.[26]

Importantly, dominance was not seen when mice (C3H.SW female) were primed with the dominant and nondominant antigens on separate APCs (i.e., a mixture of C3H.SW male cells + C57BL/6 female cells).[26] These data indicate that dominance results from competition for the APC surface between anti-B6^{dom}1 and anti-HY CTLs.

TCR/Peptide/MHC Interactions

Efficient T-cell triggering requires sustained signaling, and is correlated with the number of TCRs engaged, this being a function of epitope density and of the kinetics of TCR/peptide/MHC interactions.[45-47] The influence of epitope density on T-cell activation is straightforward because the number of triggered TCRs is a function of the logarithm of the number of complexes offered.[46-48] However, the relation between TCR affinity and T-cell triggering is more complex since optimal affinity does not correspond to maximal affinity but rather to "intermediate" affinity. Indeed, whereas each epitope with an optimal off-rate can trigger several TCRs serially, high-affinity ligands, such as anti-CD3 antibodies, are inefficient at T-cell triggering because their incapacity to dissociate does not allow serial triggering.[45]

B6^{dom}1/D^b complexes have recently been shown to display the characteristics of optimal T-cell ligands (refs. 26, 38, 49). The number of B6^{dom}1 copies on C57BL/6 splenocytes is calculated to be ~10^3/cell. This places B6^{dom}1 among the most abundant class I-restricted epitopes.[50] In addition, B6^{dom}1/D^b complexes are recognized by TCRs whose relative binding affinity is intermediate. In contrast, H-Y/D^b complexes are low density ligands (calculated to be ~10 copies per cell) recognized by high affinity TCRs. H-Y/D^b complexes should therefore be less effective at triggering many TCRs. The difference in epitope density may be explained, at least in part, by the finding that the D^b-binding affinity of B6^{dom}1 is greater than that of HY/D^b.[49]

Proposed Model

Based on the findings summarized in the preceding paragraph, we assume that B6^{dom}1 should promote a more efficient activation of cognate CTLs than HY/D^b. As a direct consequence, anti-B6^{dom}1 CTLs would expand more rapidly and swiftly outnumber anti-HY/D^b CTLs (Fig. 7.1). This would not influence CTL responses when both minor H antigens are presented on separate APCs. However, when both minor H antigens are presented on the same APC, the numerical supremacy of B6^{dom}1-specific CTLs should enable them to compete more successfully for occupancy of the APC surface. Furthermore, as CD8$^+$ CTLs can receive help only when they bind the same APC as CD4$^+$ helper cells,[18] B6^{dom}1-specific CTLs would carry off all the help available. Thus, as a consequence of their superior TCR signaling ability, dominant epitopes could inhibit T-cell responses toward nondominant epitopes.

Immunologic Role and Therapeutic Implications of Immunodominance

A Wise Strategy?

According to information discussed above, dominant antigens are more efficient T-cell ligands and trigger more potent in vivo responses than nondominant antigens. Considering

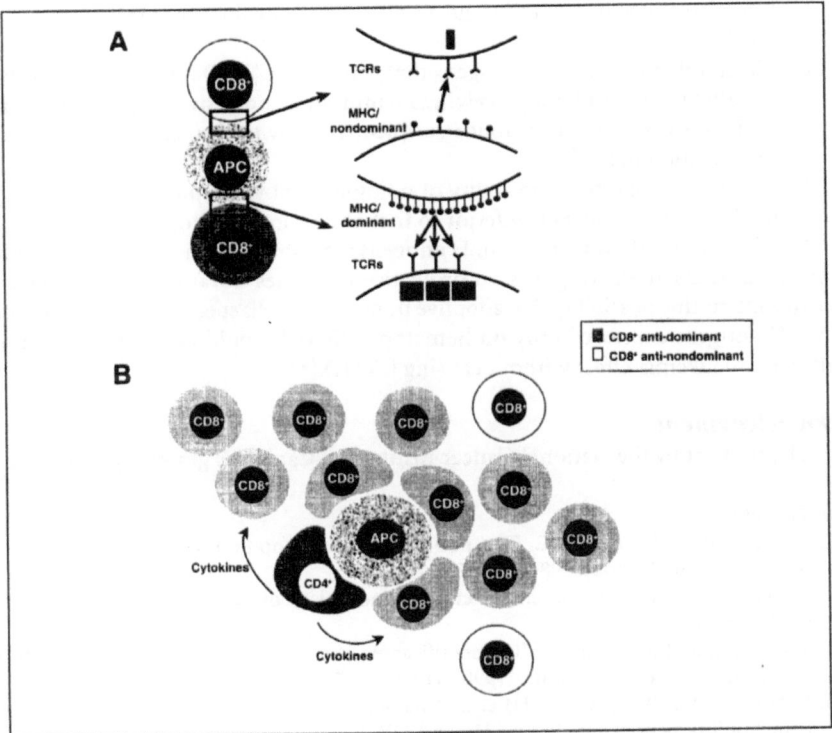

Fig. 7.1. The mechanism of immunodominance; a working model based on comparisons between B6^{dom}1 and HY minor H antigens. (A) B6^{dom}1 is more efficient than HY at T-cell triggering because B6^{dom}1 epitopes are more abundant, and each B6^{dom}1 epitope can serially engage multiple TCRs. (B) B6^{dom}1 promotes a more rapid expansion of cognate CTL clones. The numerical supremacy of anti-B6^{dom}1 CTL clones enables these clones to compete more successfully for occupancy of the APC surface.

that TCRs are degenerate ligands, and that T-lymphocytes responding to non self-antigens can crossreact with self-peptides thereby causing autoimmune diseases, the immunodominance phenomenon may in fact represent an astute "low risk/high efficiency" strategy selected by the immune system.[15,51,52] Indeed, restricting the diversity of the immune response limits the potential for autoimmune recognition, while focusing on the best epitopes confers good chances to rapidly eliminate pathogens.

Implications in Transplantation

If GVHD is elicited by a few dominant minor H antigens, prospective typing for this select group of minor H antigens could be used to improve donor selection.[53] This endeavor may be possible considering the low polymorphism of minor H antigens. However, this approach would be useful only for a limited group of patients: those with more than one MHC-matched donor. As a more general solution, the possibility to induce tolerance to dominant minor H antigens should be explored.[54,55]

If dominant minor H antigens are avoided, will nondominant minors trigger GVHD? Although nondominant minor H antigens can trigger CTLs detectable by in vitro assays, three findings suggest that their ability to trigger GVHD is negligible:

1. When presented one at a time, no a single nondominant minor H antigen can trigger GVHD;[36,37]

2. The number of minor H antigens that can cause GVHD must be very low because in both humans and mice receiving no immunosuppressive drugs, in some cases no GVHD was observed in donor/recipient pairs with numerous minor H antigen incompatibilities;[56,57]

3. In a different context, the ability of nondominant viral epitopes to elicit protective in vivo CTL responses is inferior to that of dominant epitopes.[58]

Finally, immunodominant minor H antigens are also more efficient than nondominant minor H antigens at eliciting in vivo protective responses towards leukemic cells.[59] This raises the interesting possibility that adoptive transfer of T cells specific for immunodominant minor H antigens expressed only on hematopoietic cells could be used to potentiate the graft-versus-leukemia effect without causing GVHD.[7,27]

Acknowledgment

The support by the National Cancer Institute of Canada is gratefully mentioned.

References

1. Loveland B, and Simpson E. The non-MHC transplantation antigens: Neither weak nor minor. Immunol Today 1986; 7(7&8):223-229.

2. Perreault C, Décary F, Brochu S et al. Minor histocompatibility antigens. Blood 1990; 76(7):1269-1280.

3. Wallny HJ, and Rammensee HG. Identification of classical minor histocompatibility antigen as cell-derived peptide. Nature 1990; 343(6255):275-278.

4. Rotzschke O, Falk K, Wallny HJ et al. Characterization of naturally occurring minor histocompatibility peptides including H4 and HY. Science 1990; 249(4966):283-7.

5. Fischer Lindahl K. Minor histocompatibility antigens. Trends Genet 1991; 7(7):219-24.

6. Roopenian DC. What are minor histocompatibility loci? A new look at an old question. Immunol Today 1992; 13(1):7-10.

7. Goulmy E. Minor histocompatibility antigens: From T-cell recognition to peptide identification. Hum Immunol 1997; 54(1):8-14.

8. Korngold R, Sprent J. Lethal GVHD across minor histocompatibility barriers: Nature of the effector cells and role of the H2 complex. Immunol Rev 1983; 71:5-29.

9. Rammensee HG, Klein J. Polymorphism of minor histocompatibility genes in wild mice. Immunogenetics 1983; 17(6):637-47.

10. Roopenian DC, Christianson GJ, Davis AP et al. The genetic origin of minor histocompatibility antigens. Immunogenetics 1993; 38(2):131-40.

11. den Haan J, Bontrop RE, Pool J et al. Conservation of minor histocompatibility antigens between human and non-human primates. Eur J Immunol 1996; 26(11):2680-5.

12. Kent-First MG, Maffitt M, Muallem A et al. Gene sequence and evolutionary conservation of human SMCY. Nat Genet 1996; 14(2):128-9.

13. Goulmy E. Human minor histocompatibility antigens. Curr Opin Immunol 1996; 8(1):75-81.

14. Sercarz EE, Lehmann PV, Ametani A et al. Dominance and crypticity of T-cell antigenic determinants. Annu Rev Immunol 1993; 11:729-66.

15. Deng Y, Yewdell JW, Eisenlohr LC et al. MHC affinity, peptide liberation, T-cell repertoire, and immunodominance all contribute to the paucity of MHC class I-restricted peptides recognized by antiviral CTL. J Immunol 1997; 158(4):1507-15.

16. Vitiello A, Yuan L, Chesnut RW et al. Immunodominance analysis of CTL responses to influenza PR8 virus reveals two new dominant and subdominant Kb-restricted epitopes. J Immunol 1996; 157(12):5555-62.

17. Roopenian DC, Davis AP, Christianson GJ et al. The functional basis of minor histocompatibility loci. J Immunol 1993; 151(9):4595-605.

18. Stuhler G, Walden P. Collaboration of helper and cytotoxic T-lymphocytes. Eur J Immunol 1993; 23(9):2279-86.
19. Bailey DW. Four approaches to estimating number of histocompatibility loci. Transplant Proc 1970; 2(1):32-8.
20. Wettstein PJ, Bailey DW. Immunodominance in the immune response to "multiple" histocompatibility antigens. Immunogenetics 1982; 16(1):47-58.
21. Wettstein PJ. Immunodominance in the T-cell response to multiple non-H2 histocompatibility antigens. II. Observation of a hierarchy among dominant antigens. Immunogenetics 1986; 24(1):24-31.
22. Yin L, Poirier G, Neth O et al. Few peptides dominate cytotoxic T-lymphocyte responses to single and multiple minor histocompatibility antigens. Int Immunol 1993; 5(9):1003-9.
23. Tremblay N, Fontaine P, Perreault C. T-lymphocyte responses to multiple minor histocompatibility antigens generate both self-major histocompatibility complex-restricted and cross-reactive cytotoxic T-lymphocytes. Transplantation 1994; 58(1):59-67.
24. Franksson L, Petersson M, Kiessling R et al. Immunization against tumor and minor histocompatibility antigens by eluted cellular peptides loaded on antigen processing defective cells. Eur J Immunol 1993; 23(10):2606-13.
25. Wolpert E, Franksson L, Karre K. Dominant and cryptic antigens in the MHC class I restricted T-cell response across a complex minor histocompatibility barrier: Analysis and mapping by elution of cellular peptides. Int Immunol 1995; 7(6):919-28.
26. Pion S, Fontaine P, Desaulniers M et al. On the mechanisms of immunodominance in cytotoxic T-lymphocyte responses to minor histocompatibility antigens. Eur J Immunol 1997; 27(2):421-30.
27. Perreault C, Roy DC, Fortin C. Immunodominant minor histocompatibility antigens: The major ones. Immunol Today 1998; 69-74.
28. Silins SL, Cross SM, Elliott SL et al. Development of Epstein-Barr virus-specific memory T cell receptor clonotypes in acute infectious mononucleosis. J Exp Med 1996; 184(5):1815-24.
29. Steven NM, Leese AM, Annels NE et al. Epitope focusing in the primary cytotoxic T-cell response to Epstein-Barr virus and its relationship to T-cell memory. J Exp Med 1996; 184(5):1801-13.
30. Dudley ME, and Roopenian DC. Loss of a unique tumor antigen by cytotoxic T-lymphocyte immunoselection from a 3-methylcholanthrene-induced mouse sarcoma reveals secondary unique and shared antigens. J Exp Med 1996; 184(2):441-7.
31. Huang AY, Gulden PH, Woods AS et al. The immunodominant major histocompatibility complex class I-restricted antigen of a murine colon tumor derives from an endogenous retroviral gene product. Proc Natl Acad Sci USA 1996; 93(18):9730-5.
32. Johnston JV, Malacko AR, Mizuno MT et al. B7-CD28 costimulation unveils the hierarchy of tumor epitopes recognized by major histocompatibility complex class I-restricted CD8+ cytolytic T-lymphocytes. J Exp Med 1996; 183(3):791-800.
33. Brochu S, Baron C, Hétu F et al. Oligoclonal expansion of CTLs directed against a restricted number of dominant minor histocompatibility antigens in hemopoietic chimeras. J Immunol 1995; 155(11):5104-14.
34. Berger MA, Korngold R. Immunodominant CD4+ T-cell receptor Vbeta repertoires involved in graft-versus-host disease responses to minor histocompatibility antigens. J Immunol 1997; 159(1):77-85.
35. Liu X, Chesnokova V, Forman SJ et al. Molecular analysis of T-cell receptor repertoire in bone marrow transplant recipients: Evidence for oligoclonal T-cell expansion in graft-versus-host disease lesions. Blood 1996; 87(7):3032-44.
36. Blazar BR, Roopenian DC, Taylor PA et al. Lack of GVHD across classical, single minor histocompatibility (miH) locus barriers in mice. Transplantation 1996; 61(4):619-24.
37. Korngold R, Leighton C, Mobraaten LE et al. Inter-strain graft-vs.-host disease T-cell responses to immunodominant minor histocompatibility antigens. Biol Blood Marrow Transplant 1997; 3(2):57-64.

38. Perreault C, Jutras J, Roy DC et al. Identification of an immunodominant mouse minor histocompatibility antigen (MiHA). T-cell response to a single dominant MiHA causes graft-versus-host disease. J Clin Invest 1996; 98(3):622-8.
39. Goulmy E, Schipper R, Pool J et al. Mismatches of minor histocompatibility antigens between HLA-identical donors and recipients and the development of graft-versus-host disease after bone marrow transplantation. N Engl J Med 1996; 334(5):281-5.
40. Wettstein PJ. Murine minor histocompatibility antigens detected by helper T cells. Recognition of an endogenous peptide. J Immunol 1995; 154(5):2134-43.
41. Griem P, Wallny HJ, Falk K et al. Uneven tissue distribution of minor histocompatibility proteins versus peptides is caused by MHC expression. Cell 1991; 65(4):633-40.
42. Korngold R, Wettstein PJ. Immunodominance in the graft-vs-host disease T-cell response to minor histocompatibility antigens. J Immunol 1990; 145(12):4079-88.
43. Greenfield A, Scott D, Pennisi D et al. An H-YD[b] epitope is encoded by a novel mouse Y chromosome gene. Nat Genet 1996; 14(4):474-8.
44. Wettstein PJ. Immunodominance in the T-cell response to multiple non-H2 histocompatibility antigens. III. Single histocompatibility antigens dominate the male antigen. J Immunol 1986; 137(7):2073-9.
45. Viola A, Lanzavecchia A. T-cell activation determined by T cell receptor number and tunable thresholds. Science 1996; 273(5271):104-6.
46. Valitutti S, Lanzavecchia A. Serial triggering of TCRs: A basis for the sensitivity and specificity of antigen recognition. Immunol Today 1997; 18(6):299-304.
47. Bachmann MF, Oxenius A, Speiser DE et al. Peptide-induced T-cell receptor downregulation on naive T cells predicts agonist/partial agonist properties and strictly correlates with T cell activation. Eur J Immunol 1997; 27(9):2195-203.
48. Valitutti S, Muller S, Cella M et al. Serial triggering of many T-cell receptors by a few peptide-MHC complexes. Nature 1995; 375(6527):148-51.
49. Pion S, Christianson GJ, Fontaine P et al. Shaping the repertoire of cytotoxic T-lymphocyte (CTL) responses. The molecular basis of the immunodominance effect whereby CTLs specific for immunodominant antigens prevent recognition of short non dominant antigens. Blood 1999; 952-62.
50. Levitsky V, Zhang QJ, Levitskaya J et al. The life span of major histocompatibility complex-peptide complexes influences the efficiency of presentation and immunogenicity of two clas I-restricted cytotoxic T-lymphocyte epitopes in the Epstein-Barr virus nuclear antigen 4. J Exp Med 1996; 183(3):915-26.
51. Wucherpfennig KW, Strominger JL. Molecular mimicry in T cell-mediated autoimmunity: Viral peptides activate human T-cell clones specific for myelin basic protein. Cell 1995; 80(5):695-705.
52. Barnaba V. Viruses, hidden self-epitopes and autoimmunity. Immunol Rev 1996; 152:47-66.
53. Martin PJ. How much benefit can be expected from matching for minor antigens in allogeneic marrow transplantation? Bone Marrow Transplant 1997; 20(2):97-100.
54. Cobbold SP, Adams E, Marshall SE et al. Mechanisms of peripheral tolerance and suppression induced by monoclonal antibodies to CD4 and CD8. Immunol Rev 1996; 149:5-33.
55. Davies JD, Leong LY, Mellor A et al. T cell suppression in transplantation tolerance through linked recognition. J Immunol 1996; 156(10):3602-7.
56. Martin PJ. Increased disparity for minor histocompatibility antigens as a potential cause of increased GVHD risk in marrow transplantation from unrelated donors compared with related donors. Bone Marrow Transplant 1991; 8(3):217-23.
57. Fontaine P, Langlais J, Perreault C. Evaluation of in vitro cytotoxic T-lymphocyte assays as a predictive test for the occurrence of graft vs host disease. Immunogenetics 1991; 34(4):222-6.
58. Oukka M, Manuguerra JC, Livaditis N et al. Protection against lethal viral infection by vaccination with nonimmunodominant peptides. J Immunol 1996; 157(7):3039-45.
59. Pion S, Fontaine P, Baron C et al. Immunodominant minor histocompatibility antigens expressed by mouse leukemic cells can serve as effective targets for T-cell immunotherapy. J Clin Invest 1995; 95(4):1561-8.

CHAPTER 8

Identification of Human Minor Histocompatability Antigens: Towards Clinical Benefits

Els Goulmy

Minor histocompatibility (H) antigens are readily studied in the HLA identical Bone Marrow Transplantation (BMT) setting. BMT in combination with chemoradiotherapy is used as a treatment for severe aplastic anemia, leukemia, and other hematologic malignancies. The ideal transplant situation is when BM donor and recipient have identical MHC antigens. The results of clinical BMT reveal that the selection of such MHC identical donors does not guarantee avoidance of two of the major drawbacks of allogeneic BMT, GVHD and leukemic relapse. GVHD occurs, depending on the age of the recipient and the amount of T-cell depletion of the graft, in 15-35% of HLA genotypically identical donor/recipient pairs.[1,2] T cell depletion of the donor marrow inoculum shows a reduction in the incidence and severity of GVHD but coincides with an increase of leukemia relapse. So the presence of some mature T cells in the donor bone marrow inoculum is essential for graft acceptance. T cells are also responsible for GVHD and most probably the beneficial Graft-versus-Leukemia (GVL) effect. Several clinical studies indicate a direct relationship between the GVL effect and acute and chronic GVHD.[3-5] In syngeneic BMT between identical twins and in recipients of autologous BMT, relapse rates are high. No MHC or minor H antigen disparities exist and thus no alloreactivity can be induced. In contrast, in allogeneic BMT the relapse rates are significantly lower and a relationship is seen between the GVL effect and acute and chronic GVH.[6] Since the GVH and GVL activities occur in allogeneic, HLA identical BMT, alloreactive donor T cells are likely to play an important role in these activities. In the HLA identical situation, these alloreactive donor T cells appear to be directed at the patient's disparate minor H antigens.

Human Minor H Antigens: From the Bedside to the Bench

The first report on "alloimmunity to human minor H antigen" concerned the induction of MHC restricted cytotoxic T cells (CTLs) against the male specific minor H antigen H-Y.[7] An in vitro cellular immune response against H-Y was detected in multitransfused female patient who received, after ATG pre-treatment, the bone marrow of a male HLA-identical sibling. A "take" of the marrow could not be detected.[8] Although it is still debatable whether minor H antigens induce an antibody response, donor-specific HLA-A2 restricted H-Y specific cytotoxic T lymphocytes (CTLs) and antibodies were detected.[9]

A clinical case encoded by autosomal chromosomes also opened our eyes to a possible involvement of minor H antigens in the development of GVHD in man. The occurrence of

Minor Histocompatibility Antigens: From the Laboratory to the Clinic, edited by Derry Roopenian and Elizabeth Simpson. ©2000 Landes Bioscience.

a severe GVHD in a bone marrow-transplanted male AML patient prompted us to investigate the in vitro cytotoxic activity of the patient's posttransplantation lymphocytes. The patient had been transplanted with bone marrow from an HLA identical female sibling donor. His clinical discovery, however, was complicated by severe acute and chronic GVHD. The patient's posttransplant lymphocytes had strong cytotoxic activity against the recipient's pretransplant lymphocytes but not against the lymphocytes of his HLA identical donor.[10] This reactivity was consistent with the generation of HLA-restricted CTLs by donor cells, which were specific for minor H antigens.

From additional analysis of the patient's posttransplant CTL activities, it became apparent that the minor H antigen (which we designated HA-1) was not only present on the patient's own pretransplant cells, but could also be detected on lymphocyctes from two out of three haplo-identical siblings as well as on the lymphocytes of the parents, and also on lymphocyctes from a large number of unrelated healthy individuals. The antigen HA-1 could only be recognized by the patient's posttransplant CTL if one of the patient's HLA class I antigens was present on the target cells.[11] Consequently, HA-1 was recognized in an MHC restricted fashion, an event comparable to the recognition of H-Y. Since strong anti-minor H antigen cytotoxic activity was observed in a patient suffering from severe GVHD after HLA identical BMT, it was reasonable to assume that there might exist a correlation between both in vivo and in vitro observations. Based on this concordance, we continued our search for non-HLA antigens and their influence on the outcome of BMT. The availability of clinical material from patients following HLA-identical BMT and their donors has been essential for the generation of anti-host CTLs and T helper (Th) cells with specific activity for autosomally encoded minor H antigens. As for anti-HA-1 CTLs, the generation of minor H antigen specific T cells is based on the following assumption: posttransplant (i.e., donor) cells when sensitized against the patient's own pretransplant cells, are directed against host minor H antigen that are absent from the donor cells. Several reports have demonstrated the presence of anti-host CTLs and Th in patients suffering from GVHD after HLA genotypically identical BMT. The efforts of several investigators have led to the identification of a small number of minor H antigen recognize in a classical MHC restricted fashion (for review see ref. 12).

Minor H Antigens: Characteristics

We performed detailed analyses on the characteristics of minor H antigens generated from individuals primed in vivo by minor H antigen-mismatched bone marrow grafting or by blood transfusions. We studied the male specific H-Y antigen and five non-sexlinked minor H antigen (designated HA-1 to HA-5) at three levels, i.e., immunogenetics, tissue distribution and immunogenicity (for details see ref. 12). The results of the genetic studies and tissue expression of the male-specific H-Y and of non-Y-linked minor H antigens HA-1 to HA-5, as well as the T-cell receptor (TcR) usage of some of the minor H antigen-specific CTL clones, are summarized in Table 8.1. As can be seen, different minor H antigens can be recognized in the context of different HLA alleles, yet HLA-A2.1 is used frequently as the presenting molecule.

Immunogenetics

Phenotype frequency analyses were carried out for HA-1 to HA-5. These studies revealed that some minor H antigens, i.e., HA-1, HA-2, and HA-3, appeared frequently (69%-95%), whereas others, i.e., HA-4 and HA-5, occurred with lesser (7-16%) frequencies in the healthy population (Table 8.1). An analysis of their genetic traits demonstrated a Mendelian mode of inheritance.

Table 8.1. Characteristics of human minor H antigens H-Y and HA-1 to HA-5

Minor H Antigen	HLS Restriction molecules	Phenotype frequency(%)	Tissue distribution	TCR Usage
H-Y	Various[a]	50	Broad[b]	Variable
HA-1	A2	69	Restricted[c] + leukemic cells	Vβ6.9
HA-2	A2	95	Restricted + leukemic cells	Variable
HA-3	A1	88	Broad	n.t.[d]
HA-4	A2	16	Broad	n.t.
HA-5	A2	7	Restricted + leukemic cells	n.t.

[a] HLA-A1, A2.1, B7, and B60.
[b] Expression on all hematopoietic and nonhematopoietic cell lineages.
[c] Expression restricted to the hematopoietic cell lineage.
[d] n.t., not tested.

Tissue Distribution

CTL clones were also used to analyze functional expression (the read-out being cell-mediated lympholysis) of the minor H antigens on various tissues and cells. Differential expression was observed: some, i.e., H-Y, HA-3, and HA-4, are ubiquitously expressed, whereas the expression of other minor H antigens, i.e., HA-1 and HA-2, is limited to cells of the hematopoietic lineage only. It is important to note that minor H antigens H-Y and HA-1 and HA-5 are expressed on clonogenic normal and leukemic precursors cells as well on myeloid and lymphoid leukemic cells isolated from the peripheral blood (Table 8.1).

Immunogenicity

Three sets of data are consistent with a hierarchy in immunogenicity among minor H antigens. Firstly, CTL clones reactive to HA-1 were obtained from peripheral blood lymphocytes of 3 out of 5 patients each transplanted across a multiple and probably distinct minor H barriers. Secondly, the latter HA-1 specific CTL clones derived from these 3 unrelated patients all seemed to use an identical TCR Vβ, showed remarkable similarities within the N-D-N regions, but used distinct V and β segments (Table 8.1). The latter results may be indicative of a dominant minor H antigen-specific T-cell response occurring during the development of GVHD after BMT. Thirdly, in a retrospective study, comprising 148 bone marrow donor/recipient pairs, investigating the influence of HA-1 to HA-5 mismatching on the development of GVHD, we observed a significant correlation (p= 0.02) between an HA-1 mismatch and GVHD.[13]

Are There Major Minors?

Although the number of minor H antigens is expected to be large, probably only a limited number will fulfil the criteria (i.e., frequency, tissue distribution, immunogenicity) for being a risk factor for GVHD or rejection. The fact that a significant number of bone marrow transplants between HLA-identical sibling (with optimal immunosuppression) do not lead to GVHD, suggests a hierarchy in immunogenicity. Various factors or combinations thereof (see Table 8.2) determine the immunogenic potential of particular minor H antigens (such as HA-1) resulting in efficient activation of the immune system.

Table 8.2. Are there major minors?

- Synergistic effects of minor H antigen-specific Th-CTL
- T-cell repertoire dependence
- Peptide affinity/peptide off-rate
- Minor H proteins producing peptides in different alleles
- Production of cytokines/soluble factors
- Molecular mimicry

What Are Minor H Antigens?

Minor H antigens are naturally processed peptides of intracellular minimally polymorphic "self" proteins that associate with MHC molecules. Minor H antigens are likely to be derived from evolutionarily conserved genes with important biological functions given their expression patterns. In a transplantation setting, however, these peptides can be immunogenic and can induce cellular immune responses.

Minor H antigen specific T-cell clones (Table 8.1) have been used for the identification of the chemical nature of their antigens. Since minor H antigens are recognized by T cells in association with MHC molecules, an obvious approach is to isolate the peptide antigens from their respective MHC molecules. Thanks to the technical advances of Hunt et al,[14] the application of a microcapillary high-performance liquid chromatography (HPLC)-electrospray ionization tandem mass spectrometry enabled the detection of rare peptides among a pool of MHC-bound peptides. Our joint forces allowed the first chemical identification of human minor H antigen (Table 8.3, refs. 15-18).

Human Minor H Antigens: From Bench to Bedside

The clinical potential of minor H antigens remains to be evaluated. Nevertheless, some areas of clinical application are worth mentioning.

Utility for Diagnosis

The utility for diagnosis in BM donor selection is self-evident. We are currently identifying minor *H* genes so that typing on the molecular level can be performed. This could result in improved donor selection and will enable identification of BMT donor/recipient pairs with high risk of minor H antigen-induced GVHD.

Immunomodulation of GVHD

The pronounced immunogenicity of minor H antigen HA-1 in GVHD together with the recently obtained amino acid composition of the cognate peptide,[18] is the foundation for attempts at immunomodulation of GVHD. Designing minor H antigen peptide-analogues that function as MHC or T-cell receptor antagonists might interfere with the harmful anti-host minor H antigen-directed T-cell reactivities following HLA-identical BMT. Our nonhuman primate study[19] shows the possibility of using transgenic chimpanzees or rhesus monkeys as a model for studying BMT-related reactivities such as GVHD.

An animal model is also required for studying the potential application of minor H antigens with broad tissue distribution, such as H-Y and HA-3, in the specific induction of tolerance in minor H antigen negative BM donors to prevent GVHD and in minor H antigen negative BM recipients to prevent rejection. Achieving tolerance to specific minor H antigens

Table 8.3. Identification of human minor H antigens

Restriction Molecule	Minor H Antigen	Peptide (no. of amino acids)	Origin	Ref.
HLA-A2.1	HA-2	YIGEVLVSV (9AA)	Nonfilamentous class I myosin? involved in cell locomotion	15
HLA-B7	H-Y	SPSVDKARAEL (11AA)	SMCY; transcription factor Spermatogenesis?	16
HLA-A2.1	H-Y	FIDSYICQV (9AA)	SMCY	17
HLA-A2.1	HA-1	VLHDDLLEA (9AA)	?	18

prior to transplantation would decrease the necessity for using pharmacological immuno-suppression.

Immunotherapy for Leukemia

Most promising is immunotherapy for leukemia using CTLs specific for minor H antigen peptides for the treatment of refractory, residual, or relapsed leukemia. Minor H antigens with restricted tissue distribution i.e., expression on hematopoietic cells including leukemic progenitors and circulating leukemic cells (see Table 8.1) are candidates for adoptive immuno-therapy of leukemia. Upon transfusion, either pre-BMT as part of the conditioning regimen or post-BMT as adjuvant therapy, the minor H antigen peptide-specific CTLs could eliminate the patient's leukemia cells and, if of recipient origin, also the patient's hematopoietic cells, but will spare the patient's nonhematopoietic cells. If necessary, subsequent donor BMT could restore the patient's hematopoietic system. The ideal situation is to generate minor H antigen peptide-specific CTLs ex vivo from minor H antigen negative BM donors for minor H antigen positive patients. A universal option would be to generate "pre-fab" minor H antigen peptide-specific CTLs by the use of minor H antigen negative healthy blood donors with common HLA-homozygous haplotypes. Transduction of these CTLs with a suicide gene makes elimination of the CTL possible in case adverse effects occur. Future research should also focus on the possibility of minor H antigen donor immunization of minor H antigen negative BM donors prior to BMT to minor H antigen-positive high-risk relapse recipients.

Conclusion

Matching for the HLA antigens to improve the success of organ and bone marrow grafting revealed the existence of non-HLA or minor histocompatibility (H) antigens. Tissue transplanted between individuals with genetic identity at HLA genes results in graft rejection and graft-versus host disease (GVHD). Therefore, disparities for minor H antigens between donor and recipient constitute a potential risk for graft failure and GVHD.

For decades, minor H antigens have been regarded as disturbing entities used to account for unwanted immune reactivities in recipients of MHC-matched bone marrow transplants. Now that the molecular identity of these antigens is being discovered, we may apply this knowledge to improve the results of human bone marrow transplantation.

Acknowledgments

The human minor H antigen studies from the bedside to the bench and from the bench to the bedside summarized here resulted from combined efforts of (in alphabetical order): Astrid Bakker, Els Blokland, Ronald Bontrop, Marleen de Bueger, Cécile van Els, Fred Falkenburg, Joke den Haan, Dick van der Harst, Linda Liem, Ellen van Lochem, Tuna Mutis, Carla Reinhardus, Jon van Rood, Ellen Schrama, Jos Pool, Jaak Vossen. The biochemical identification of the human minor H antigens has been performed in collaboration with Professors Engelhard and Hunt and their collaborators from the University of Virginia, Charlottesville, USA. I am grateful to Sonja Mesander and Leny de Vries for typing the manuscript. The work was supported in part by grants from the Dutch Organization for Scientific Research (NWO), the Dutch Cancer Foundation, the Dutch Ziekenfondsraad, the J.A. Cohen institute for Radiopathology and Radiation Protection (IRS), the Macropa Foundation, the European Community for Biotechnology and the Leukemia Society of America.

References

1. Bortin MM, Horowitz Mm, Mrsic M et al. Progress in bone marrow transplantation for leukemia: A preliminary report from the Advisory Committee of the International Bone Marrow Transplant Registry. Transplant Proc 1991; 23:61-62.
2. Beatty PG, Hervé P. Immunogenetic factors relevant to acute graft-versus-host disease. In: Burakoff SJ, Deeg HJ, Ferrara S et al, eds. Graft-Versus-Host Disease: Immunology, Pathophysiology and Treatment. New York: Marcel Dekker, 1990; 12:415-423.
3. Weiden PL, Flournoy N, Thomas ED et al. Antileukemic effect of graft-versus-host disease in human recipients of allogeneic marrow grafts. N Engl J Med 1979; 300:1068-1073.
4. Weiden PL, Fournoy KM, Flournoy N et al. Antileukemic effect of chronic graft-versus-host disease. Contribution to improved survival after allogeneic marrow transplantation. N Engl J Med 1981; 304:1529-1532.
5. Butturini A, Bortin MM, Gale RP. Graft-versus leukemia following bone marrow transplantation. Bone Marrow Transplant 1987; 2:233-242.
6. Horowitz MM, Gale RP, Sondel PM et al Graft-versus leukemia reactions after bone marrow transplantation. Blood 1990; 75:555-562.
7. Goulmy E, Termijtelen A, Bradley BA et al. Alloimmunity to human H-Y. Lancet 1976; 2:1206.
8. Goulmy E, Termijtelen A, Bradley BA et al. Y-antigen killing by T cells of women is restricted by HLA. Nature 1977; 266:544-545.
9. Van Leeuwen A, Goulmy E, Van Rood JJ. Major histocompatibility complex-restricted antibody reactivity mainly, but not exclusively directed against cells from male donors. J Exp Med 1979; 150:1075-1083.
10. Goulmy E, Gratama JW, Blokland E et al. Recognition of an as yet unknown minor transplantation antigen by posttransplant lymphocytes from an AML patient. Exp Hem 10: (Suppl. 10) 1982:127-129.
11. Goulmy E, Gratama JW, Blokland E et al. A minor transplantation antigen detected by MHC restricted cytotoxic T lymphocytes during graft-versus-host disease. Nature 1983; 159:159-161.
12. Goulmy E. Human minor histocompatibility antigens: New concepts for marrow transplantation and adoptive immunotherapy. Imm Rev 1997; 157:125-140.
13. Goulmy E, Schipper R, Pool J et al. Influence of minor histocompatibility antigen mismatches on the development of graft-versus-host disease after bone marrow transplantation from HLA identical donors. N Engl J Med 1996; 334:281-285.
14. Hunt DF, Henderson RA, Shabanowicz J et al. Characterization of peptides bound to class I MHC molecule HLA-A2.1 by mass spectrometry. Science 1992; 255:1261-1263.
15. Den Haan JMM, Sherman NE, Clokland E et al. Identification of graft-versus-host disease associated human minor histocompatibility antigen. Science 1995; 268:1478-1480.

16. Wang W, Meadows LR, Den Haan JMM et al. Human H-Y: A male-specific histo-compatibility antigen derived from the SMCY protein. Science 1995; 269:1588-1590.
17. Meadows L, Wang W, Den Haan JMM et al. The H-Y antigen presented by HLA-A*0201 contains a post-translationally modified cysteine residue: A common peptide modification that significantly affects T-cell recognition. Immunity 1997; 6:273-281.
18. Den Haan JMM, Meadows LM, Wang W et al. The minor histocompatibility antigen HA-1: A diallelic gene with a single amino acid polymorphism. Science 1998; 279:1054-1057.
19. Den Haan JMM, Bontrop RE, Pool J et al. Conservation of minor histocompatibility antigens between human and non-human primates. Eur J Immunol 1996; 26:2680-2685.

CHAPTER 9

Mapping Human Minor Histocompatibility Genes

Patrick G. Beatty and James C. Jenkin

In allogeneic human blood and bone marrow transplantation, it has been apparent for many years that although disparity for HLA can contribute to both graft rejection and graft versus host disease, there are also important non-HLA loci involved.[1] This conclusion is based on observations that graft versus host disease (GVHD) is extremely rare when transplants are carried out between genotypical identical twins, but is present even when there is genetic identity for HLA by inheritance, such as would happen when transplants take place between HLA matched siblings.[2] Indeed, if no immunosuppression is given in marrow transplant between HLA matched siblings, the incidence of fatal graft versus host disease is extremely high.[3] Thus, although matching for HLA is an important, though not an absolute, requirement for a successful transplant,[4] clearly there are other antigen systems involved: by definition the minor histocompatibility (H) antigens. These minor H antigens are particularly important when transplants are carried out between HLA matched unrelated individuals. The probability of mismatch for minor H antigens is higher for an unrelated patient/donor pair than if the transplant is between HLA matched siblings, as siblings would have some shared inheritance of genes encoding minor H antigens, and unrelated patient/donor pairs would not.[5]

An effort to develop typing systems for human minor H antigens would be particularly relevant in unrelated transplantation. In sibling transplants, it is often difficult to find even one HLA matched sibling; hence, it is unlikely that there would be several from which to choose based upon minor H types. Thus, knowing about minor H antigen disparities in matched sibling transplants would be useful only to predict the probability of graft versus host disease, and perhaps devise appropriate therapeutic preventative measures. In unrelated transplants, however, although some patients can never find a full HLA-match, there are many patients that have many possible HLA-matched donors.[6] In such a setting, it would be useful to not only type these potential donors and patients for HLA, but also for all known clinically relevant minor H antigens. Thus, one could hopefully decrease the risk of graft versus host disease by choosing the best minor H antigen matched donor.[7] Development of a typing system would first involve identification of the genes encoding the minor antigens, developing a DNA based typing system, and then performing clinical correlation studies to assess relevance. Such DNA-based systems have already been developed for HLA typing.[8]

As a prelude to developing such a typing system, it is important to have a grasp of how many different minor H antigens exist in humans, and to understand under what circumstances these antigens are clinically important. For instance, in certain mouse systems, it appears that the risk of allogeneic complications is directly proportional to the number of

Minor Histocompatibility Antigens: From the Laboratory to the Clinic, edited by Derry Roopenian and Elizabeth Simpson. ©2000 Landes Bioscience.

minor H antigens for which they are disparate.[9] In other mouse systems, it appears that at any given time there are a limited number of minor H antigens that are important, but if one manipulates the strain combination such that the donor and recipient are now matched for that minor H antigen, another takes precedence: the so-called phenomenon of immunodominance.[10] By understanding the applicability of such findings to humans, it should then be possible to predict the feasibility of developing a DNA based typing system in humans.

Our approach has been to map and enumerate minor *H* loci in humans, and determine whether the repertoire of minor H antigens correlates with specific HLA molecules. In doing so, we hope it will be possible to approximate of the number of potentially relevant human minor *H* loci. If there is a limited number of such genes, the next step would be to use standard methodology for gene cloning to identify the actual genes, and then develop a DNA-based typing system.

Genomics

The system employed in our studies is based on techniques developed for positional gene cloning for genetic diseases. This approach has been successful in isolation of many genes such as that encoding the BRCA1 breast cancer gene,[11] and the long Q-T syndrome in patients with a propensity for cardiac arrhythmias.[12] The concept is to identify a large family pedigree in whom the disease phenotype has a clear segregation pattern. When such a family is identified, DNA from each family member is then genotyped with respect to thousands of molecular "markers" developed for genetic mapping. Some of these markers represent polymorphic genes, which encode known proteins, but more recently, many markers have been developed that detect non-coding polymorphisms that have no apparent relevance to cellular function, but are nonetheless very useful in segregation analyses. Thus, one can deduce the inheritance based on the pattern of polymorphism of the marker loci among family members. One then runs a computer program, which asks whether there is any segment of a chromosome (as defined by the genetic markers) whose segregation pattern within the family is the same as the segregation pattern of the disease phenotype within that family. If one or more segments are found to segregate, one would use another family with the same disease, in whom one would perform a similar segregation analysis. The statistical analysis of such gene mapping has been well described. Certain assumptions need to be made with respect to disease penetrance, homozygosity, double-crossovers, etc.

Our approach has been to apply this method to the genetic mapping of human minor *H* loci. The "disease phenotype" that we follow is the lysis of lymphoblastoid cell lines (LCL) derived from individual members of each of several already gene-mapped families by a cytotoxic T-cell clone known to react with a donor minor H antigen. One can then derive a family segregation analysis of this minor H phenotype, and perform a linkage study with respect to the thousands of known gene markers within one or more families.

Localizing Human Minor *H* Loci

A male patient was identified who suffered from chronic graft versus host disease. His original marrow donor was his HLA genotypically identical sister. Cytotoxic T-lymphocyte (CTL) clones were derived from the peripheral blood of this patient, as described elsewhere.[13,14] The initial screen of the T-cell clones was against ^{51}Cr labeled patient and donor lymphoblastoid cell lines. Clones were chosen for further analysis if they lysed patient LCL, but not donor LCL. All T-cell clones were donor derived (female), as would be anticipated in the blood of a post-marrow transplant patient. Nineteen clones met these criteria, and were further analyzed to determine the HLA restricting element of each done. A large panel of lymphoblastoid cell lines were chosen, each cell of which had one of the HLA antigens shared by patient and donor but not the others. We were then able to identify four CTL clones which appeared to be restricted by HLA-B7, as each lysed some, but not all, of fourteen HLA-B7 positive cells, but failed to lyse cells that did not express HLA-B7. Three different patterns of reactivity against HLA-B7 positive cells were noted and, hence, we hypothesized that three minor H antigens were being presented in the peptide binding groove of HLA-B7 on the patient LCL-targets.

The Eccles Institute of Human Genetics, University of Utah, in collaboration with the Centre d'Etude du Polymorphisme Humain (CEPH, Paris), maintains LCL cell lines derived from forty families that have been saturation mapped with thousands of polymorphic genetic markers. Many of these families had been in previous disease association studies. Each family has, at a minimum, four grandparents, two parents, and several siblings. We transfected the HLA-B7 molecule into representatives of several of these families, then tested them for lysis by our T-cell clones, to establish that the minor H antigen detected by that clone was indeed expressed in that family. In order for our T-cell clone to lyse a specific LCL, both the minor H antigen and the appropriate HLA restricting element must be present (in this case, HLA-B7). For two of our CTL clones, we were able to identify two and three families, respectively, in which the respective minor H antigen appeared to be expressed, that is, some LCL from each family were lysed by the clone. We then transfected LCLs from every family member with our HLA-B7 construct and, hence, rendered each LCL informative for expression of relevant minor H antigens. CTL clone MD2 was tested against LCL from two families (1331 and 1362), as shown in Figures 9.1 A and B. Informative segregation was established for each of the two families, and the segregation pattern was then submitted for computer analysis, asking if there was a unique genomic localization correlating to these two particular segregation patterns. As shown in Figure 9.2, such a region was identified with logarithm of the difference (LOD) scores in this region with many loci above 3 some with 4 to 5. LOD scores above three are considered statistically significant. A LOD score of 5 indicates that the correlation of individuals expressing the minor H antigen with a given marker locus would have a change of $< 1/10^5$. (The variation of LOD scores within the tested region represents differences in the informativeness of specific markers.) The boundaries of the genomic region in question are set by known recombinations in this particular section of chromosome 22. Thus, with the two families tested with this CTL clone, this segment of the genome segregated precisely as did the segregation of lysis patterns observed in these two families. As there were no further recombinations within this genomic region in these two families, we cannot further refine where within this region the gene in question might lie. Such an analysis would require other approaches. For instance, we could study additional families, and hope for a recombination within our target region. Alternatively, we could use somatic cell immunoselection techniques to produce minor H antigen-loss variant LCLs that carry deletions and other informative mutations.[15] Finally, we could apply other genomic approaches, such as screening yeast artificial chromosomes (YACs) from this genetic region for the minor *H* gene[16] by transfection into minor H antigen-negative HLA-B7$^+$ transformed cells, and, hence, further narrow in on the minor *H* genes.

A second clone (NF3) was mapped in a similar fashion, as shown for three families in Figures 9.3 A-C. Again, informative segregation was found in each family, and upon computer analysis, a localization was made to 58 cM portion of chromosome 11 (Fig. 9.4).

We have thus shown that this patient had circulating in his blood CTLs capable of recognizing two distinct minor H antigens in the context of HLA-B7. The genes encoding these minor H antigens localize to distinct portions of the genome (Fig. 9.5). From our preliminary analysis, there is likely at least one more minor antigen restricted by HLA-B7 that is present in the blood of this individual. We are currently studying whether the other

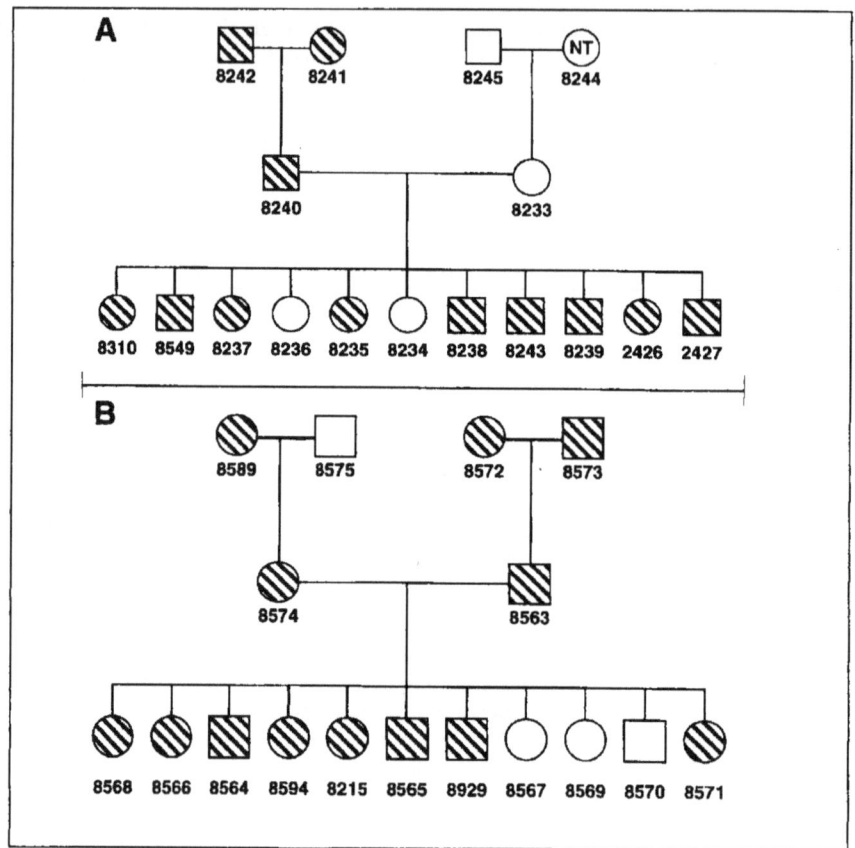

Fig. 9.1. (A) Summary of CTL data for family K1331. Shaded circles (females) or squares (males) indicate strong lysis ("positive phenotype,") open symbols indicate minimal lysis ("negative phenotype.") (B) Summary of CTL data for family K1362. Shaded circles or squares indicate strong lysis, open symbols indicate minimal lysis. NT: not tested.

HLA antigens expressed by this patient (A3, A29, DR2, and DR4) might also be serving as restricting elements for peripheral T cells which may also be recognizing minor H antigens.

We are also in the process of studying other bone marrow transplantation patients who express HLA-B7 to determine whether their T cells recognize minor H antigens whose encoding genes localize to the same portions of chromosomes 11 and 22 or rather, to other parts of the genome. If they localize to the same portions, we can not be certain that they are encoded by the same genes. However, if they localize to different parts of the genome, we can conclude that the CTL clones are recognizing distinct minor H antigens.

The genetic region encoding minor H antigens that we have identified were based on the activity of isolated T cells circulating in the patient's blood while undergoing a GVH reaction. This does not prove that T cells reacting to the antigens encoded by these minor H loci were causative of the GVHD process. However, the testing of this possibility should be feasible by combining our genetic localizations with the complementary analyses of Ginsburg and Nichols (Chapter 11) in which they are searching for genetic regions that correlate with GVHD.

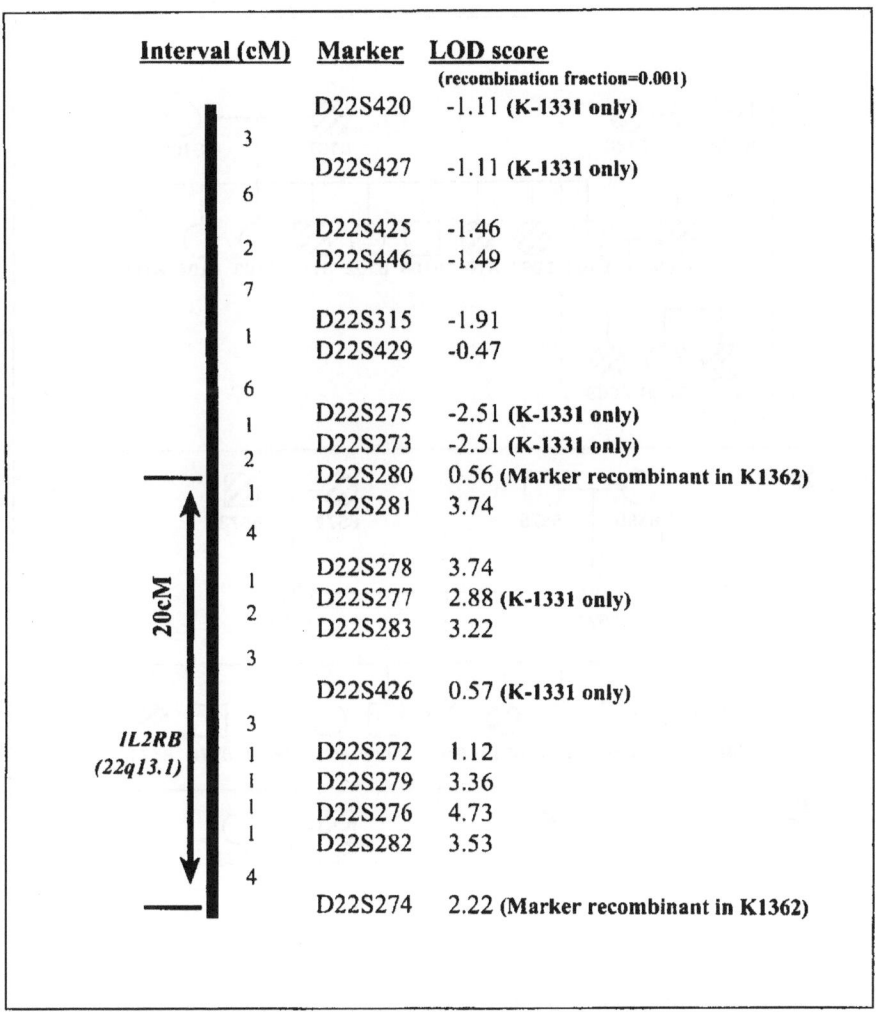

Fig. 9.2. Genetic map of human chromosome 22. Pairwise LOD scores are given for polymorphic markers and the postulated minor *H* locus in two transfected CEPH families, K-1362 and K-1331. IL2RB = interleukin 2 receptor, beta.

Conclusion

We have identified two genetic regions encoding minor H antigen expressed in the context of HLA-B7. The extension of these studies should allow a enumeration of the minimal set of human minor antigens presented by HLA-B7, and by other HLA molecules. If there are a small number loci that appear to be of particular significance, it would be relatively straightforward, although potentially laborious, to clone these genes, and develop molecular reagents necessary for DNA typing.

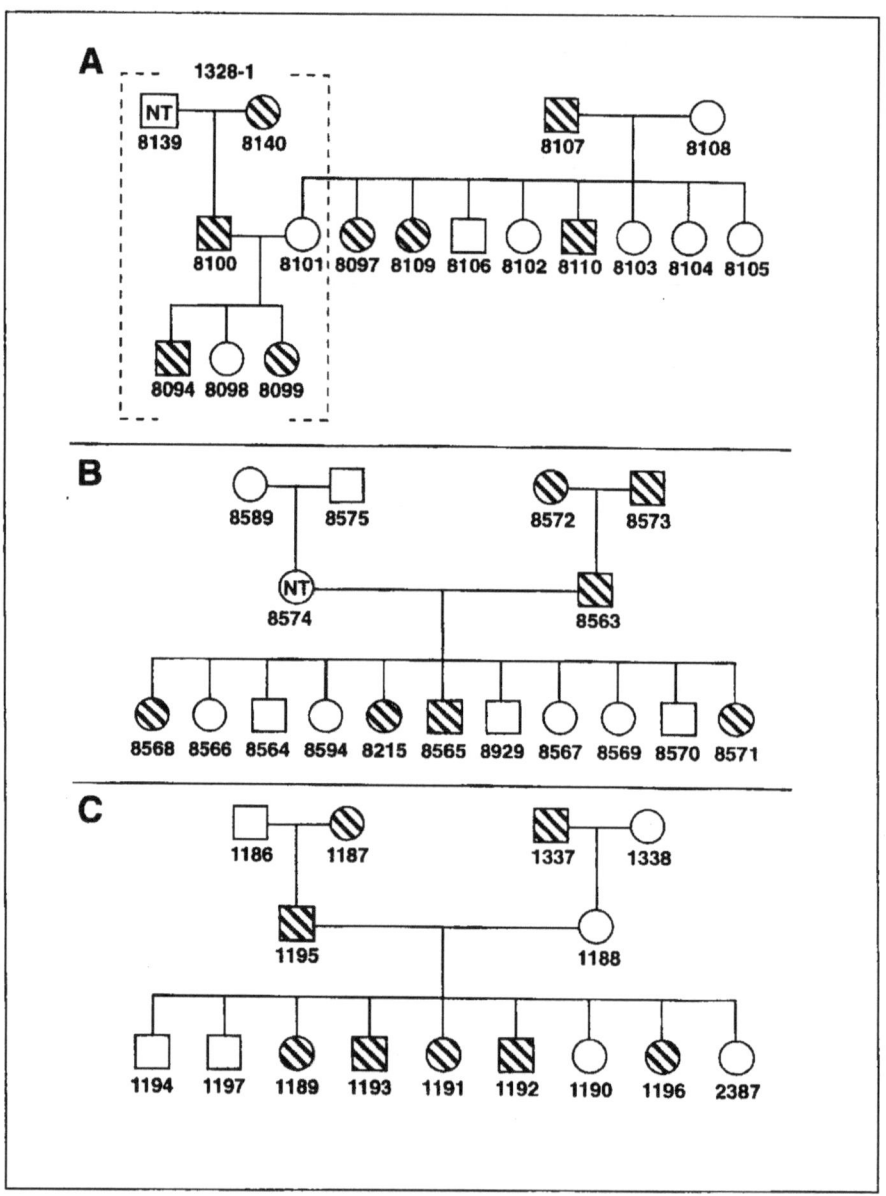

Fig. 9.3. (A) Summary of CTL data for family 1328. Shaded circles (females) or squares (males) indicate strong lysis; open symbols indicate minimal lysis. NT indicates cell was unable to be transfected and tested. (B) Summary of CTL data for family 1362. Shaded circles (females) or squares (males) indicate strong lysis (positive phenotype); open symbols indicate minimal lysis (negative phenotype). (C) A summary of CTL data for family 1416. Shaded circles (females) or squares (males) indicate strong lysis; open symbols indicate minimal lysis. LOD scores are shown for polymorphic markers and the postulated minor H locus in three transfected CEPH families:1328, 1362, and 1416.

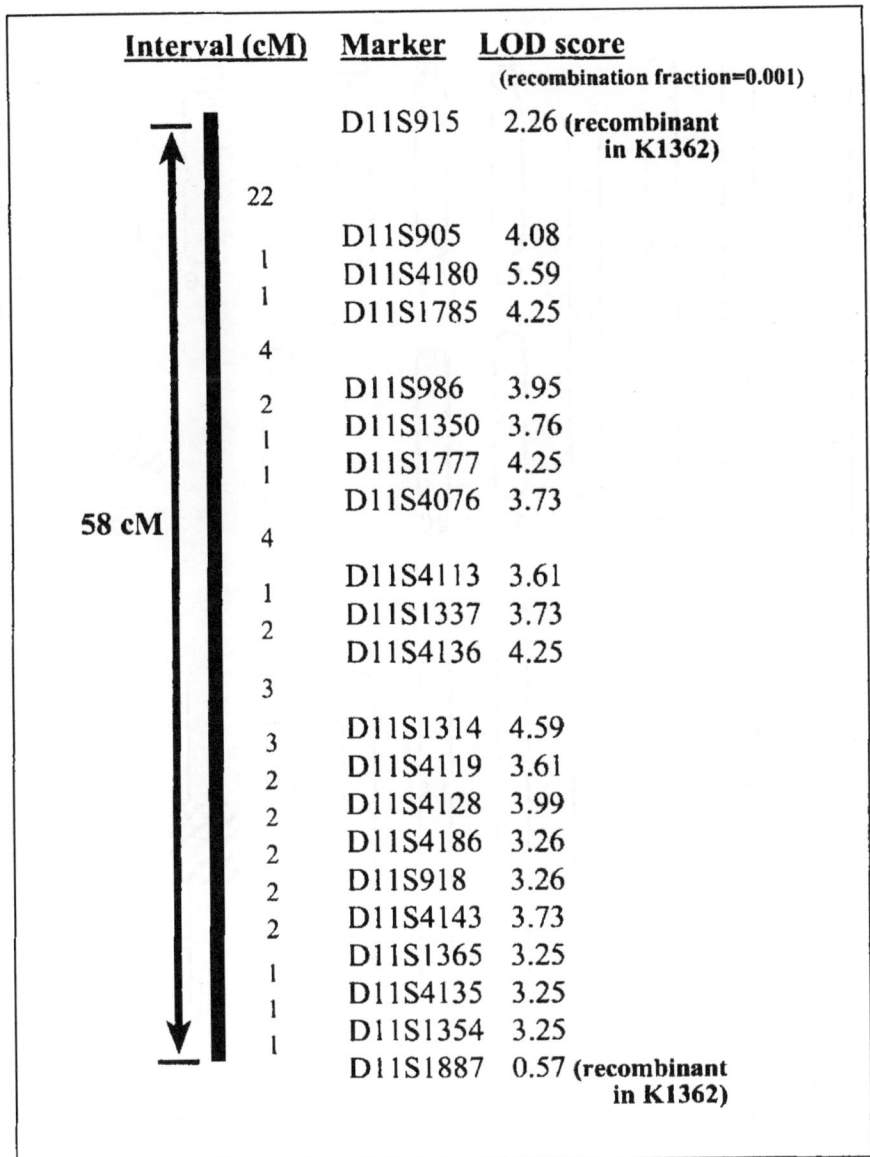

Fig. 9.4. Genetic map of part of human chromosome 11. Pairwise LOD scores are given for polymorphic markers.

References

1. Beatty PG. Minor histocompatibility antigens. Exp Hematol 1993: 21:1514-1516.
2. Beatty PG. Role of histocompatibility antigens in the development of graft-versus-host disease. In: Graft-Versus-Host Disease. Burakoff SJ, Deeg HJ, Ferrara J et al, eds. 2nd ed. New York: Marcel Decker, 1996: 607-614.

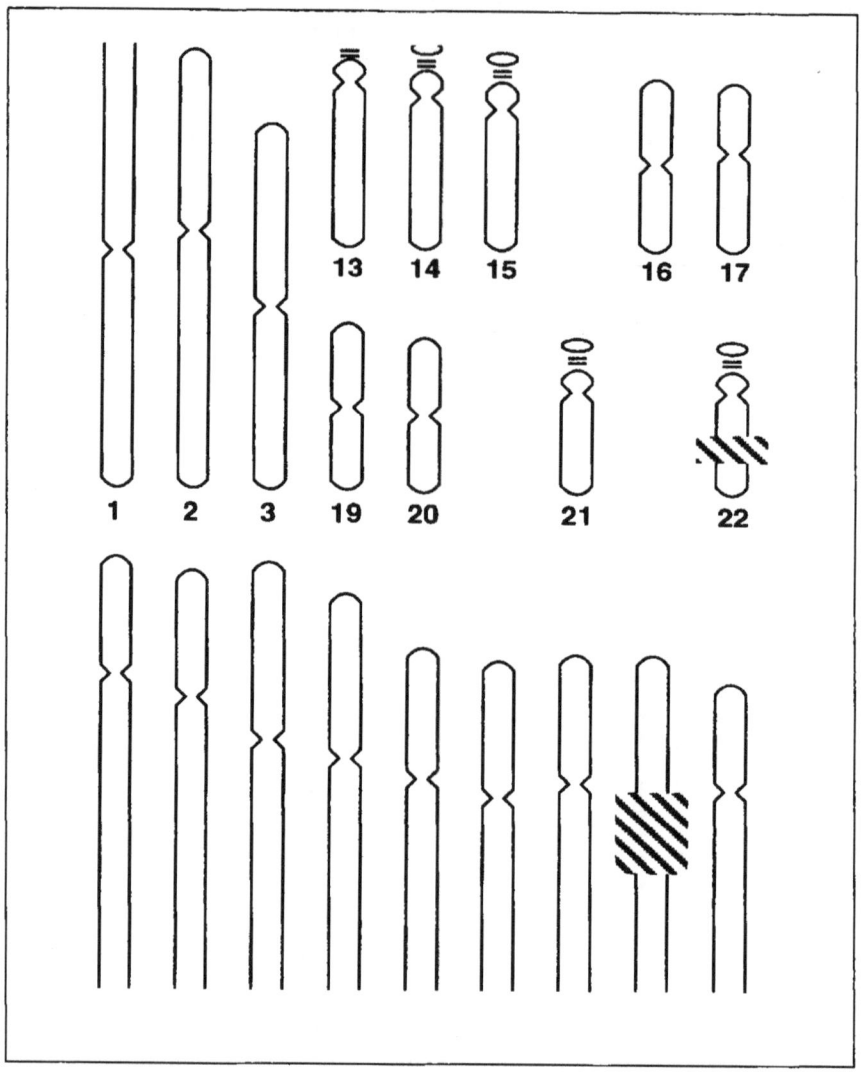

Fig. 9.5. Depiction of the human genome with localization of genes encoding two minor H antigens to chromosomes 11 and 22.

3. Sullivan KM, JH Deeg, Sanders J et al. Hyperacute graft vs. host disease in patients not given immunosuppression after allogeneic marrow transplantation. Blood 1986; 67:1172-1175.

4. Beatty PG: The immunogenetics of bone marrow transplantation. Transfusion Medicine Reviews 1994; 8:45-58.

5. Martin PJ. Increased disparity for Minor-HA as a potential cause of increased GVHD risk in marrow transplantation from unrelated donors compared to related donors. Bone Marrow Transplantation 1991; 8:217-223.

6. Beatty PG, Mori M, Milford E. Impact of racial genetic polymorphism on the probability of finding an HLA-matched donor. Transplantation 1995; 60:778-783.

7. Beatty PG. National Heart, Lung, Blood Institute Workshop on the Importance of Minor Histocompatibility Antigens in Marrow Transplantation. September 16-17, 1996; Bethesda, Maryland. Exp Hemat 1997; 25:548-558.

8. Beatty PG. The molecular revolution in histocompatibility testing: Relevance to blood and marrow transplantation. In: Garratty G, ed. Applications of molecular biology. Bethesda: American Association of Blood Banks, 1997:51-72.

9. Loveland B, Simpson E. The non-MHC transplantation antigens: Neither weak nor minor. Immunol Today 1986; 7:223-228.

10. Korngold R, Wettstein PJ. Immunodominance in the graft-versus-host disease T-cell response to minor histocompatibility antigens. J Immunol 1990; 145:4079-4088.

11. Goldgar DE, Cannon-Albright LA, Oliphant A et al. Chromosome 17q linkage studies of 18 Utah breast cancer kindreds. Am J Hum Genet 1993; 52:743-748.

12. Spawski I, Tiomothy KW, Vincent GM et al. Molecular basis of the long-QT syndrome associated with deafness. N Engl J Med 1997; 336:1562-7.

13. Gubarev MI, Jenkin JC, Leppert MF et al. Localization to chromosome 22 of a gene encoding a human minor histocompatibility antigen. J Immunol 1996; 157:5448-5454.

14. Gubarev MI, Jenkin JC, Otterrud BE et al. Localization to chromosome 11 of a gene encoding a human minor histocompatibility antigen. Exp Hemat 1998; 26: 978-981.

15. Zuberi AR, Dudley ME, Christianson GJ et al. 1994. Gene mapping in a murine cell line with cytotoxic T-cell immunoslection. Genomics 19; 334-340.

16. Zuberi AR, Christianson GJ, Dave SB et al. 1998. Expression screening of overlapping yeast artificial chromosomes identifies a clone that carries the mouse H3a minor histocompatibility antigen gene. J Immunol 1998; 161:821-828.

CHAPTER 10

Applicability of Matching for Minor Histocompatibility Antigens in Human Bone Marrow Transplantation

Paul J. Martin

Marrow transplantation is often complicated by graft-versus-host disease (GVHD). Billingham[1] formulated the requirements for GVHD as (1) genetically determined histocompatibility differences between the recipient and donor, (2) the presence of immunocompetent cells in the graft, and (3) inability of the host to react against or reject the graft. When there is HLA-disparity between the donor and recipient, GVHD is initiated by donor T cells that recognize recipient HLA antigens that are not shared by the donor.[2] For this reason, HLA compatibility between the donor and recipient represents the overriding concern in donor selection for allogeneic marrow transplantation. When the donor and recipient are HLA-identical siblings, GVHD is initiated by donor T cells that recognize recipient minor histocompatibility (H) antigens that are not shared by the donor.[3]

Only 30-35% of the patients who could benefit from marrow transplantation have an HLA-identical sibling who can serve as a donor, and the opportunity for selecting among multiple HLA-identical siblings occurs infrequently. The ability to identify minor H antigens is therefore not likely to be useful for optimizing selection of an HLA-matched related donor. An assessment of minor H antigen disparity, however, would allow more judicious use of interventions to prevent GVHD. Complications caused by immunosuppressive medications or by removal of T cells from the graft could be avoided when these measures are not actually needed.

Transplantation from an HLA-identical unrelated donor is an alternative for patients who lack a donor in the family.[2] This approach has been made feasible by the development of large registries of HLA-typed individuals willing to serve as marrow donors. With unrelated transplantation, there is often a choice among multiple potential donors, particularly for individuals with common HLA types. The ability to identify minor H antigens would therefore be of considerable use in optimizing the selection of unrelated donors for marrow transplantation.

The feasibility of typing and matching for minor H antigens in unrelated marrow transplantation depends on the number of loci that encode antigens that can initiate GVHD, the number of alleles at each locus, and the distribution of alleles in the population. With few possible exceptions,[4-6] the polymorphism and allele distribution of human minor H antigens have not been defined, and the loci that encode antigens capable of causing GVHD

Minor Histocompatibility Antigens: From the Laboratory to the Clinic, edited by Derry Roopenian and Elizabeth Simpson. ©2000 Landes Bioscience.

have not been identified. Thus the feasibility of typing and matching for minor H antigens in unrelated marrow transplantation remains unknown.

Rationale for Development of a Seven Locus Model

Formula that describe the probability of minor antigen disparity in a marrow transplant recipient can be derived from fundamental principles of population genetics.[7] Given that most of the more than 40 minor *H* loci in mice are likely to have only two known alleles,[8] we postulated that human *H* loci have similarly limited polymorphism. If it is further assumed that minor H antigens are codominantly expressed and that the alleles of minor H loci are in Hardy-Weinberg equilibrium with gene frequencies p and q, then the probability of recipient disparity at any given locus can be expressed as pq (1-0.5 pq) when the donor and recipient are siblings and as 2pq (1-pq) when the donor and recipient are unrelated. Regardless of the exact allele frequencies, the probability of disparity at any given locus is always greater for unrelated pairs than for siblings.

We postulated that recipient disparity at a single locus is sufficient to cause GVHD if the graft contains T cells and if no posttransplant immunosuppression is given. The number of minor *H* loci encoding antigens capable of causing GVHD in humans can be estimated from collective data indicating that 17 of 63 recipients (27%) transplanted with unmodified (T cell-replete) marrow from an HLA-identical sibling did not develop GVHD when no post transplant immunosuppression was given.[9-11] The absence of GVHD in these recipients suggests that disparity for minor H antigens was absent. Although the allele distribution for minor *H* loci is not known, we used a 75/25 distribution as a representative average and calculated that the probability of recipient disparity at each locus with this distribution [pq (1-0.5 pq)] is 0.17. The corresponding probability of compatibility at each locus is 0.83, and the probability of compatibility at each of 7 loci (.83[7]) is 0.27. In this way, we estimated that in any individual, approximately 7 loci encode minor H antigens that can cause GVHD.

To explore this model further, we assumed that all loci encoding minor antigens that cause GVHD have approximately equivalent biologic activity in contributing to the disease and that recipient disparities at multiple loci have additive effects in contributing to GVHD severity. From the binomial distribution, we calculated that 68% of unrelated pairs and 34% of sibling pairs have $\geq$ 2 disparities (Table 10.1). These estimates are remarkably similar to the proportions of patients who experience clinically significant grades II-IV acute GVHD after transplantation of unmodified (T cell-replete) marrow from HLA-identical donors when methotrexate and cyclosporine are used for prophylaxis.[12] In the same models, 36% of unrelated pairs and 10% of sibling pairs have $\geq$ 3 disparities, similar to the proportions of patients who experience the more severe grades III-IV GVHD. As a corollary of this model, it follows that a single disparity is not sufficient to cause GVHD when methotrexate and cyclosporine are given for prophylaxis. Although the models predict that approximately 7 minor *H* loci encode antigens that can cause GVHD in any given individual, the total number of such loci in the population is likely to be much larger because individual allelic types of MHC molecules present distinctive peptides that function as minor H antigens.

Further Evaluation of a Seven Locus Model

The estimate that approximately 7 minor *H* loci might encode peptides antigens that can cause GVHD poses a paradox because most individuals are HLA-heterozygous and have two different alleles at each of the highly polymorphic loci in the MHC.[13] If HLA-C and HLA-DP are taken together with HLA-A, HLA-B, HLA-DR and HLA-DQ, and if hybrid HLA-DQα and DQβ pairing is considered,[14-15] then most individuals have 10-14 different HLA molecules potentially capable of presenting peptides as minor H antigens. This disparity between the estimated number of loci encoding minor H antigens that cause

Table 10.1. Seven locus model for minor antigens that cause GVHD

	Donor Relationship	Loci Typed	Degree of Disparity*			
			0	**≥ 1**	**≥ 2**	**≥ 3**
1	Unrelated	0	.08	.92	.68	.36
2	Unrelated	3	.23	.77	.36	.09
3	Unrelated	4	.25	.75	.34	.08
4	Sibling	0	.27	.75	.34	.10
5	Sibling	1	.33	.67	.27	.06

*Results in lines 1 and 4 show the expected proportions of recipients with the indicated degrees of disparity when donors are selected without typing and matching for the alleles of any locus. Lines 2 and 5 show the expected proportions of recipients with the indicated degrees of disparity when donors are selected after typing and matching for the number of loci designated. Line 3 shows the expected proportions of recipients with the indicated degrees of disparity after selecting the best match among 5 donors tested at 4 loci. Calculations were based on the assumption that all loci have a uniform 75/25 allele distribution.

GVHD and the number of MHC molecules that could present minor antigens could be explained if certain types of MHC molecules (e.g., HLA-B35) do not present peptides that cause GVHD because of limitations in the repertoire of polymorphic peptides encoded by minor H loci. If this scenario were correct, certain HLA alleles should be associated with a relatively reduced risk of GVHD, while others should be associated with a relatively increased risk.

Associations between individual HLA antigens and risk of GVHD have been reported previously,[16-19] but results between different studies have not been consistent. With rigorous statistical methods to correct for multiple comparisons, we were unable to find convincing evidence for any such association in a study population of more than 2,500 recipients transplanted with marrow from an HLA-identical sibling.[25] These results suggest that the variety of minor H antigens is not constrained by limitations in the repertoire of polymorphic peptides collectively encoded by minor H loci. Moreover, the lack of heterogeneity among the antigens encoded at any locus made it impossible to identify the classes of HLA molecules capable of presenting polymorphic peptides as minor H antigens that cause GVHD.

Although the MHC molecules encoded within any given locus showed no heterogeneity in their association with GVHD, different classes of MHC molecules could show considerable heterogeneity in their ability to present peptides that cause GVHD. Thus the disparity between the estimated number of loci encoding minor antigens that cause GVHD and the number of MHC molecules that could present minor antigens could be explained if certain classes of MHC molecules (e.g., HLA-C or HLA-DP) do not present peptides that cause GVHD. In this model, the variety of minor H antigens in an individual is constrained by limitations in the diversity of MHC molecules and not by limitations in the repertoire of polymorphic peptides encoded by minor H loci. If this scenario were correct, then the variety of minor H antigens should be substantially lower in HLA-homozygous individuals than in HLA-heterozygous individuals, and HLA-homozygous individuals should have a substantially lower risk of GVHD than HLA-heterozygous individuals. Contrary to this expectation, we found that HLA-homozygous individuals and HLA-heterozygous individuals have comparable risks of GVHD after transplantation from an HLA-identical sibling.[20]

The unexpectedly high incidence of GVHD in HLA-homozygous individuals might be explained in two ways. First, the T-cell receptor repertoire is likely to be wider in HLA-homozygous than in HLA-heterozygous individuals because of differences in the extent of negative selection during T-cell development in the thymus.[21] Second, the determinant density for any given minor antigen is likely to be greater with HLA-homozygous cells than with HLA-heterozygous cells. For certain minor H antigens, the number of peptide-MHC complexes in HLA-heterozygous recipient antigen-presenting cells might not reach the threshold of T-cell receptor ligation required for triggering a substantial clonal response among donor T cells, but the same minor H antigen could conceivably initiate GVHD when the number of peptide-MHC complexes is increased two-fold in HLA-homozygous recipient antigen presenting cells, thereby surpassing the threshold of T-cell receptor ligation required for activation. If certain peptide-MHC complexes can function as minor H antigens on HLA-homozygous antigen presenting cells but not on HLA-heterozygous antigen presenting cells, then the variety of minor antigens expressed by the two types of cells might be similar, despite the differences in diversity of MHC molecules.

Rationale for Development of a Thirteen Locus Model

The variety of minor H antigens involved in GVHD appears not to be constrained either by limitations in the repertoire of polymorphic peptides or by limitations in the diversity of MHC molecules. For this reason, we considered the possibility that more than 7 minor H loci in any given individual might encode peptide antigens that can cause GVHD. As a first step, we evaluated a model incorporating the assumption that the number of loci encoding minor H antigens capable of causing GVHD is approximately equivalent to the number of different MHC molecules in any individual. In a model with 13 loci in which a 75/25 allele distribution is taken as a representative average, 33% of sibling pairs have no disparity or have disparity at a single locus (Table 10.2). This proportion fits with the proportion of recipients who did not develop GVHD when no immunosuppression was given after marrow transplantation from an HLA-identical sibling (27%).

In the model with 7 loci, we assumed that recipient disparity at a single locus is sufficient to cause GVHD if no posttransplant immunosuppression is given. The model with 13 loci predicts that recipient disparity at a single locus is not sufficient to cause GVHD when no posttransplant immunosuppression is given. This prediction is consistent with observations that disparity at a single minor *H* locus is not sufficient to cause GVHD in mice.[22,23] From the binomial distribution, a 13 locus model predicts that 81% of unrelated pairs and 38% of sibling pairs have ≥ 3 disparities (Table 10. 2). These estimates are similar to the proportions of patients who experienced grades II-IV acute GVHD after transplantation of unmodified (T cell-replete) marrow from HLA-identical donors when methotrexate and cyclosporine are used for prophylaxis.[12] As a corollary of this model, it follows that disparity at 2 loci is not sufficient to cause GVHD when methotrexate and cyclosporine are given for prophylaxis.

Potential Benefits of Typing and Matching

We have evaluated the potential benefits of typing and matching for individual minor H antigens with both the 7 locus model[7,24] and the 13 locus model. In the 7 locus model, 67% of sibling pairs would be expected to have ≥ 2 disparities if one locus was typed and

Table 10.2. Thirteen locus model for minor antigens that cause GVHD

	Donor Relationship	Loci Typed	Degree of Disparity*		
			0-1	≥ 2	≥ 3
1	Unrelated	0	.06	.94	.81
2	Unrelated	6	.32	.68	.36
3	Unrelated	12	.26	.74	.38
4	Sibling	0	.33	.67	.38
5	Sibling	1	.37	.63	.33

*Results in lines 1 and 4 show the expected proportions of recipients with the indicated degrees of disparity when donors are selected without typing and matching for the alleles of any locus. Lines 2 and 5 show the expected proportions of recipients with the indicated degrees of disparity when donors are selected after typing and matching for the number of loci designated. Line 3 shows the expected proportions of recipients with the indicated degrees of disparity after selecting the best match among 5 donors tested at 12 loci. Calculations were based on the assumption that all loci have a uniform 75/25 allele distribution.

found to be mismatched, and 27% of sibling pairs would be expected to have ≥ 2 disparities among the six untyped loci if one locus was typed and found to be matched. In comparing the group known to be mismatched at the tested locus with the group known to matched at the tested locus, the odds ratio for the probability of having ≥ 2 disparities is 5.49. In the 13 locus model, 63% of sibling pairs would be expected to have ≥ 3 disparities if one locus was typed and found to be mismatched, and 33% of sibling pairs would be expected to have ≥ 3 disparities among the 12 untyped loci if one locus was typed and found to be matched. In comparing the group known to be mismatched at the tested locus with the group known to be matched at the tested locus, the odds ratio for the probability of having ≥ 3 disparities is 3.4. The odds ratios from both models fit well with data reported by Goulmy et al.[6] In comparing HA-1 minor antigen-mismatched sibling pairs with HA-1-matched sibling pairs, the odds ratio for grades II-IV GVHD was 5.4 (95% C.I., 1.0-56).

With the 7 locus models, typing and matching at a single locus decreased the proportion of sibling pairs with ≥ 2 disparities from 34% to 27% (Table 10.1), and with the 13 locus model, typing and matching at a single locus decreased the proportion of sibling pairs with ≥ 3 disparities from 38% to 33% (Table 10.2). Results from both models are consistent with the limited benefits of typing and matching for the *HA-1* locus as reported by Goulmy et al.[6] In their study, the overall incidence of grades II-IV GVHD was 55%, and the incidence of grades II-IV GVHD among HA-1-matched pairs was 51%. All considerations lead to the conclusion that typing and matching at a single locus will not substantially decrease the risk of GVHD.

The potential value of typing and matching for minor H antigens in unrelated marrow transplantation was estimated by comparing the distribution of GVHD disparities for untyped antigens against the benchmark distribution of GVHD disparities for sibling transplantation without typing and matching.[7] If the distributions of disparity are similar, then the risk of GVHD should be similar, again assuming that minor antigens involved in GVHD have equivalent biologic activity in contributing to the disease and assuming that recipient disparities at multiple loci have additive effects in contributing to GVHD severity. When 3 loci are typed and matched in the 7 locus model, the distribution of disparities for the remaining 4 untyped loci in unrelated pairs approximates the distribution of disparities in sibling pairs without typing and matching (Table 10.1). When 6 loci are typed and matched in the 13 locus model, the distribution of disparities for the remaining 7 untyped loci approximates the distribution of disparities in sibling pairs without typing and matching (Table 10.2). These results suggest that typing and matching for approximately half of the minor antigens that contribute to GVHD would decrease the incidence and severity of GVHD in unrelated transplantation to the level currently expected for sibling transplantation.

In further analysis of the 7 locus model, we estimated the potential benefit of testing multiple minor H antigens in unrelated transplantation.[24] If it were possible to select the best match among five unrelated donors by testing the alleles at 4 of the 7 minor H loci involved in GVHD, the distribution of disparities would approximate the benchmark distribution of disparities for sibling recipients without typing for minor H antigens (Table 10.1). A similar analysis was carried out with the 13 locus model to determine how many loci would have to be tested so that the proportion of recipients with ≥ 3 disparities achieved by selecting the best match among 5 unrelated donors would approximate the benchmark proportion of recipients with ≥ 3 of disparities for sibling recipients without typing for minor H antigens. Results of this analysis showed that it would be necessary to have assays that could detect disparity at 12 of the 13 loci encoding minor antigens capable of causing GVHD (Table 10.2).

Conclusion

Although there is little benefit to be gained by typing and matching at a single locus, substantial benefit would be possible if multiple loci could be typed. The success of matching would be enhanced and the requirement for testing large numbers of donors would be mitigated by having assays that would allow typing and matching for most of the loci encoding antigens that could cause GVHD in a specific patient. The task of typing and matching, however, will be complicated by the likelihood that individual types of MHC molecules present distinctive peptides that function as minor H antigens. If each individual type of MHC molecule were able to present, on average, only one or perhaps two polymorphic peptides as minor antigens that can cause GVHD, then the task of typing and matching might be much less daunting than it would seem at first glance. Our understanding of the magnitude of this task would be greatly enhanced by better information about the number of minor H antigens and the relevant MHC restriction specificities involved in causing GVHD.

References

1. Billingham RE. The biology of graft-versus-host reactions. Harvey Lect. New York: Academic Press, 1966; 62:21-78.
2. Anasetti C, Hansen J. Bone marrow transplantation from HLA-partially matched related donors and unrelated volunteer donors. In: Forman SJ, Blume KG, Thomas ED, eds. Bone Marrow Transplantation. Boston:Blackwell Scientific Publications 1994; 665-679.
3. Perreault C, Decary F, Brochu S et al. Minor histocompatibility antigens. Blood 1990; 76:1269-1280.
4. den Haan JMM, Sherman NE, Blokland E et al. Identification of a graft versus host disease-associated human minor histocompatibility antigen. Science 1995; 268:1476-1480.
5. Wang W, Meadows LR, den Haan JM et al. Human H-Y: A male-specific histocompatibility antigen derived from the SMCY protein. Science 1995; 269:1588-1590.

6. Goulmy E, Schipper R, Pool J et al. Mismatches of minor histocompatibility antigens between HLA-identical donors and recipients and the development of graft-versus-host disease after bone marrow transplantation. New Engl J Med 1996; 334:281-285.

7. Martin PJ. Increased disparity for minor histocompatibility antigens as a potential cause of increased GVHD risk in marrow transplantation from unrelated donors compared with related donors. Bone Marrow Transplant 1991; 8:217-223.

8. Graff RJ. Histocompatibility systems, except H2: Mouse. In: Altman PA, Katz DD, eds. Inbred and genetically defined strains of laboratory animals. Part 1, mouse and rat. Bethesda, MD: Federation of American Societies for Experimental Biology, 1979:118-120.

9. Lazarus HM, Coccia PF, Herzig RH et al. Incidence of acute graft-versus-host disease with and without methotrexate prophylaxis in allogeneic bone marrow transplant patients. Blood 1984; 64:215-220.

10. Elfenbein G, Graham-Pole J, Weiner R et al. Consequences of no prophylaxis for acute graft-versus-host disease after HLA identical bone marrow transplantation. Blood 1987; 70:305a (abstract).

11. Sullivan KM, Deeg HJ, Sanders J et al. Hyperacute GVHD in patients not given immunosuppression after allogeneic marrow transplantation. Blood 1986; 67:1172-1175.

12. Beatty PG, Hansen JA, Longton G et al. Marrow transplantation from HLA-matched unrelated donors for treatment of hematologic malignancies. Transplantation 1991; 51:443-447.

13. Dupont B, Yang SY. Histocompatibility. In: Forman ST, Blume KG, Thomas ED, eds. Bone Marrow Transplantation. Boston: Blackwell Scientific Publications, 1994:22-40.

14. Charron DJ, Lotteau V, Turmel P. Hybrid HLA-DC antigens provide molecular evidence for gene transcomplementation. Nature 1984; 312:157-159.

15. Giles RC, DeMars R, Chang CC et al. Allelic polymorphism and transassociation of molecules encoded by the HLA-DQ subregion. Proc Natl Acad Sci USA 1985; 82:1776-1780.

16. Storb R, Prentice RL, Hansen JA et al. Association between HLA-B antigens and acute graft-versus-host disease. The Lancet 1983; II:816-819.

17. Bross DS, Tutschka PJ, Farmer ER et al. Predictive factors for acute graft-versus-host disease in patients transplanted with HLA-identical bone marrow. Blood 1984; 63:1265-1270.

18. Weisdorf D, Hakke R, Blazar B et al. Risk factors for acute graft-versus-host disease in histocompatible donor bone marrow transplantation. Transplantation 1991: 51:1197-1203.

19. Smyth LA, Witt CS, Christiansen FT et al. The MHC influences acute graft-versus-host disease in MHC matched adults undergoing allogeneic bone marrow transplantation. Bone Marrow Transplant 1993:12:351-355.

20. Martin PJ, Petersdorf EW, Anasetti C et al. HLA homozygosity and the risk of graft-versus-host disease. Tissue Antigens 1997; 50:119-123.

21. van Meerwijk JP, Marguerat S, Lees RK et al. Quantitative impact of thymic clonal deletion on the T-cell repertoire. J Exp Med 1997; 185:377-383.

22. Blazar BR, Roopenian DC, Taylor PA et al. Lack of GVHD across classical, single minor histocompatibility (miH) locus barriers in mice. Transplant 1996; 61:619-624.

23. Korngold R, Leighton C, Mobraaten LE et al. Interstrain graft-vs.-host disease T-cell responses to immunodominant minor histocompatibility antigens. Biol Blood & Marrow Transplant 1997; 3:57-64.

24. Martin PJ. How much benefit can be expected from matching for minor antigens in allogeneic marrow transplantation? Bone Marrow Transplant 1997; 20:97-100.

25. Martin PJ, Gooley T, Anasetti C et al. HLAs and risk of acute graft-vs.-host disease after marrow transplantation from HLA-identical siblings. Biol Blood Marrow Transpl 1998; 4:128-133.

CHAPTER 11

Genetic Linkage Analysis to Identify Minor Histocompatibility Loci Contributing to Graft-Versus-Host Disease

David Ginsburg and William C. Nichols

Graft-versus-host disease (GVHD), a major complication of clinical bone marrow transplantation (BMT), results from the recognition of determinants in the host as foreign by immunologic effector cells derived from the transplanted donor bone marrow.[1,2] Though not generally thought of in these terms, GVHD is a genetic disease in that it is determined entirely by genetic differences between the host and recipient. Syngeneic bone marrow transplantation is not associated with significant GVHD, whereas GVHD of increasing severity is observed with greater genetic between host and donor. In the clinical application of BMT, HLA identical siblings are generally the preferred source of donor marrow.[3] GVHD arising in this setting is presumably due to genetic disparity at loci encoding minor histocompatibility (H) antigens[4-6]. As reviewed elsewhere in this book, minor H genes are generally thought to represent simple polymorphic amino acid sequences presented as small peptides in the context of a specific MHC molecules.

Candidate Human Minor H Genes Contributing to GVHD

Though immunologic approaches have identified epitopes recognized by cytotoxic T-cell clones derived from patients with GVHD, the clinical significance of these determinants for GVHD remains unknown. HY was the first human minor H antigen to be characterized at the genetic level. Polymorphic peptide sequences recognized in the context of two class I alleles have been identified within the SMCY protein sequence.[7] However, HY appears to play only a minor role in the development of GVHD in humans.

In 1996, Behar et al,[8] reported the first association of a polymorphism within a known human gene with a significant increase in risk for GVHD. These workers identified a leucine to valine polymorphism in codon 125 of the PECAM-1 (CD31) gene. Fourteen HLA identical BMT recipient/donor pairs in which the recipient developed grade III/IV GVHD and 32 with grade 0 GVHD were analyzed. A mismatch for the CD31 polymorphism between the recipient and donor was associated with a markedly increased risk of acute GVHD.

Goulmy and coworkers[5] have characterized a panel of 5 minor H antigens, four restricted by the HLA-A2 class I allele, and one by HLA-A1. They have termed these minor antigens HA-1 through HA-5. Typing lymphocytes for the presence or absence of these antigens by reactivity with the corresponding cytotoxic T-cell clone has demonstrated that

Minor Histocompatibility Antigens: From the Laboratory to the Clinic, edited by Derry Roopenian and Elizabeth Simpson. ©2000 Landes Bioscience.

they are inherited as simple Mendelian traits in an autosomal codominant fashion. Purification and microsequencing of the peptide corresponding to HA-2 identified this antigen as being derived from a novel member of the myosin heavy chain gene family.[9] Goulmy and coworkers[10] also reported analysis of the potential contribution of HA-1 to HA-5 to the risk for GVHD in HLA identical BMT. This analysis suggests an important role for HA-1, with minimal or no significant contributions from the other HA antigens. HA-1 has recently been identified as a nonapeptide derived from an allele of KIAA0223, a gene defined by a cDNA of unknown function, cloned from an acute myelogenous leukemia.[11]

Using a novel genetic approach, Beatty and coworkers[12] recently mapped the locus for an HLA-B7-restricted minor *H* gene defined by a cytotoxic T-cell clone isolated in their laboratory. Lymphoblastoid cell lines derived from individuals in several large pedigrees were phenotyped for this minor H antigen in a cellular cytoxicity assay after transfection with a B7 expression plasmid. These pedigrees had already been extensively studied with a large panel of polymorphic DNA markers. By this approach, these investigators were able to map this minor *H* gene to the long arm of chromosome 22.

Genome Scan to Map *GVHD* Genes

Our laboratory has sought to identify human minor *H* genes by a positional cloning approach using a novel type of sib-pair analysis. This method requires no assumptions about the biologic function of these *GVHD* genes or their mode of inheritance, only that they are genetically determined. The general strategy is illustrated in Figure 11.1. DNA samples are collected from a large panel of BMT patients and their HLA identical sibling donors. These specimens are genotyped for an extensive set of polymorphic DNA markers spaced throughout the human genome. The full set of genotyping data is then analyzed for evidence of an association for any specific marker with the occurrence of GVHD.

The possible patterns of inheritance for a putative *GVHD* gene and a nearby genetic marker are illustrated schematically in Figure 11.2. In the ideal case, a highly polymorphic genetic marker will distinguish between each of the parental alleles (indicated as A-D in Figure 11.2). In a region of the genome devoid of *GVHD* genes, two siblings would be expected to have inherited identical alleles from the corresponding parental segments in approximately 25% of pairs, completely unmatched segments in 25% of sibling pairs, and would share a single allele in the remaining 50% of pairs, regardless of GVHD status. In contrast, a mismatch for an minor H antigen between the patient and donor would be expected to increase the risk of GVHD. Thus, a polymorphic DNA marker in a nearby segment of DNA close to the minor *H* gene would also more often be mismatched in patient/donor pairs in which GVHD had occurred. Thus, the analysis consists essentially of evaluating the degree of allele sharing between the patient and sibling donor for the full panel of polymorphic DNA markers, looking for significantly less allele sharing than that expected by chance in GVHD sibling pairs and excess sharing of alleles in pairs without GVHD. This approach requires no assumptions about the biologic functions of these potential *GVHD* genes and is not dependent on any specific model for inheritance, though we might anticipate a codominant pattern of inheritance, as for the major histocompatibility locus.

In the initial phases of our analysis, a series of grade III/IV GVHD sibling pairs and a series of grade 0 GVHD sibling pairs were genotyped and statistical analysis performed to compare the degree of allele sharing between the two groups. More recently we have restricted our studies to only grade III/IV GVHD patients. For comparison, the random distribution of allele sharing that would be expected in an unselected group of patients is calculated by a computer algorithm. Although focusing our study on only patients with grade III or IV GVHD has significantly limited the number of available samples, there are several reasons for this choice. First, it is likely that the penetrance of GVHD in sibling pairs

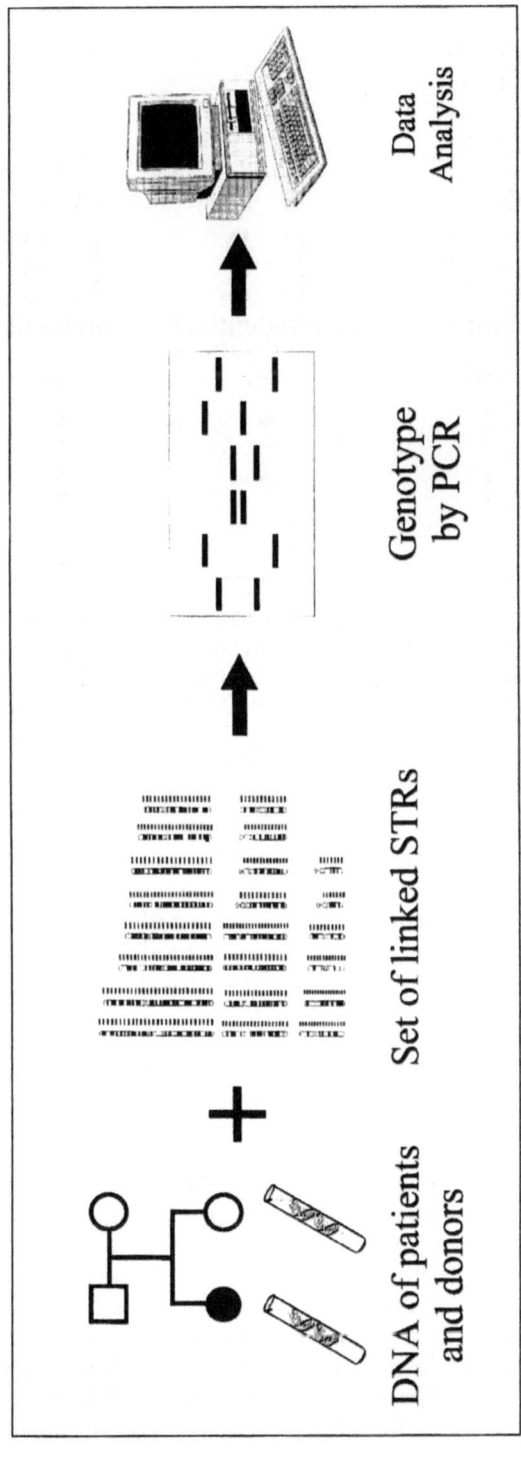

Fig. 11.1. Overview of genetic approach to the identification of genes contributing to GVHD. DNA samples are prepared from bone marrow transplant patients and their HLA-identical sibling donors. These DNAs are analyzed by PCR to determine the genotype for a large panel of short tandem repeat DNA sequence polymorphisms (w) spaced throughout the human genome. The data are then analyzed by a complex set of computer algorithms to identify markers for which mismatch is associated with a greater risk of GVHD.

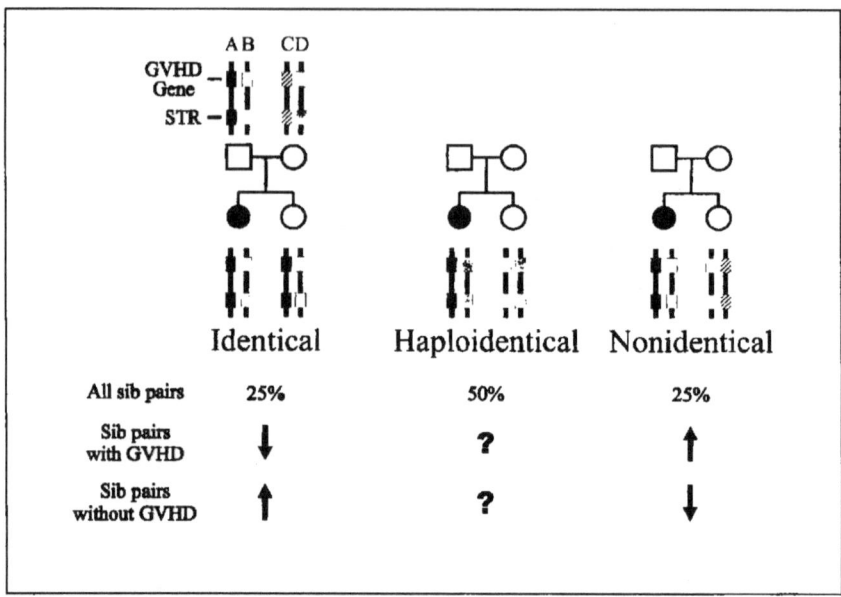

Fig. 11.2. Linkage of an STR marker to a *GVHD* Gene. The highly polymorphic marker (STR) has a different allele associated with each of the parental chromosomes (A-D). If located close to the *GVHD* gene, the STR and the gene of interest will be inherited together. By chance, 25% of siblings would be predicted to be identical at any given STR marker (50% haploidentical, and 25% non-identical). For an STR marker linked to a *GVHD* gene, less sharing of alleles would be expected among patients with severe GVHD and their siblings, whereas increased sharing would be expected in the pairs without GVHD.

who are mismatched at a minor *H* gene is incomplete. That is, only a subset of patients who are mismatched with their sibling donor may actually go on to develop significant GVHD. The absence of GVHD in some minor H antigen mismatched sibling pairs could be the result of GVHD prophylaxis immunosuppression or potential interactions with other modifying genes. Thus, the grade 0 GVHD group, in addition to recipient/donors matched for the minor *H* gene(s) of interest, undoubtedly also contains a significant number of genetically mismatched pairs.

By restricting our studies to severe GVHD, we also sought to avoid the potential misdiagnosis of GVHD that is particularly likely to occur in mild disease (grade I and II), where the manifestations can be difficult to distinguish from a number of other non-GVHD BMT complications. Thus, the grade III/IV GVHD group is likely to contain the clearest genetic information content, with all patients expected to be genetically mismatched with their donors at one or more minor *H* gene(s). In our current study, we are using a large panel of polymorphic DNA markers located throughout the human genome at an average spacing of approximately 10 cM, such that any given minor *H* gene contributing to GVHD should be located no more than 5 cM from the closest tested marker.

The polymorphic peptides containing minor H antigens are expected to be presented in the context of specific MHC molecules. Thus, it is likely that different patterns of minor *H* genes contributing to GVHD will be observed as a function of HLA type. For this reason, our current data analysis is being stratified by HLA genotype. We are initially focusing on

HLA-A2. The allele frequency of HLA-A2 is 0.25, with approximately 50% of patients thus expected to carry at least one HLA-A2 allele.

Is *CD31* a GVHD Gene?

The recent report of Behar et al. demonstrating a strong association of GVHD with recipient/donor match for a CD31 protein polymorphism[8] led us to perform a similar analysis in our much larger set of patients.[13] Patient and donor were genotyped by PCR analysis, as shown in Figure 11.3A. Patients were also stratified by HLA type as described above, with HLA-A2 genotype determined by PCR (Figure 11.3B). Table 11.1 shows the estimated frequencies of sibling pair genotype combinations expected, based on the allele frequencies we observed for each of the codon 125 alleles. The genotype combinations indicated by underlining are compatible; that is the recipient does not contain an allele foreign to the donor. For example, a donor genotype of leucine/valine though not identical, is compatible with a leucine/leucine recipient. However, the reverse—a leucine/leucine in the donor and leucine/valine in the recipient, is incompatible; in the latter case, the recipient's valine allele would be seen as foreign by the donor.

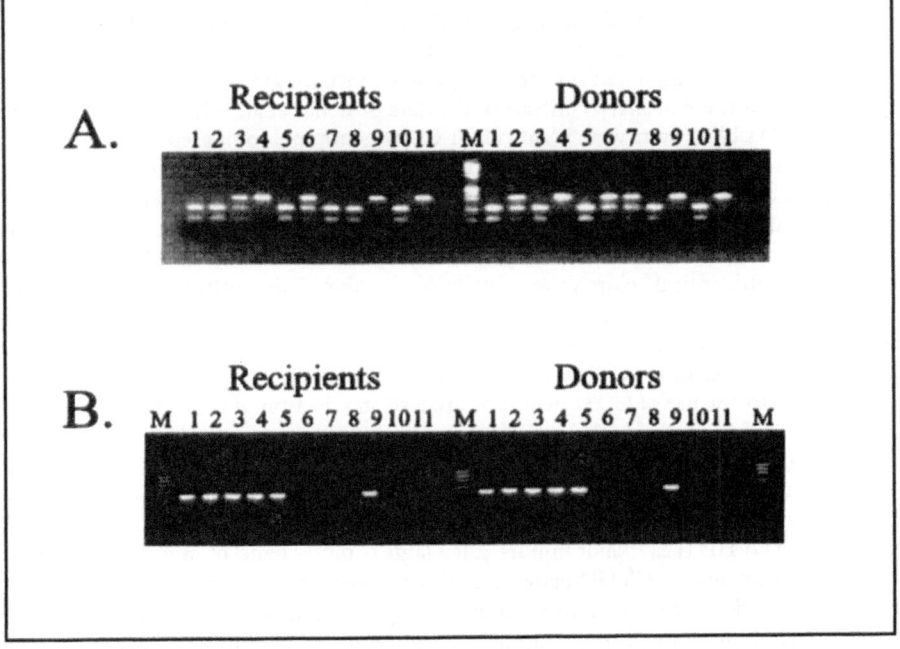

Fig. 11.3. Genotype analysis for the CD31 leucine/valine codon 125 polymorphism and HLA-A2. (A) PvuII digestion of *CD31* gene PCR products. DNA samples for 11 BMT recipients are shown to the left and their corresponding donors to the right. Lane 1 in both panels is a leucine/leucine homozygote (presence of PvuII site), recipient lane 2 is a heterozygote, and recipient lane 4 a valine/valine homozygote. M indicates a size marker lane. (B) HLA-2 PCR products of the recipient/donors show in A. The presence of a band indicates that an individual carries at least one HLA-A2 allele, whereas the absence of a band indicates that an individual carries no HLA-A2 alleles. Note that all recipients and their corresponding donors show the identical patterns. [Taken from ref. 13, with permission.]

Table 11.1. Estimated frequencies of sib pair genotype combinations

	Donor's Genotype		
	Leucine/Leucine	Leucine/Valine	Valine/Valine
Recipient's Genotype			
Leucine/Leucine	**0.1415**	0.0940	0.0156
Leucine/Valine	0.9040	**0.3125**	0.0935
Valine/Valine	0.0156	0.0935	**0.1398**

Estimated probability of identity: 0.594
Estimated probability of compatibility: 0.781
Those frequencies shown in **bold** represent identity between sib pairs, those underlined represent compatibility. [Taken from ref. 13, with permission.]

Table 11.2 shows the results of our analysis of 301 donor/recipient sibling pairs, 74 with grade 0 GVHD and 227 with grade III/IV GVHD. No significant differences were observed in the fraction of pairs identical for CD31 genotype (69% for grade 0 vs. 63% for grade III/IV) or in the percent that are compatible (85% for grade 0 and 77% for grade III/IV). These results are in sharp contrast to the data of Behar et al,[8] where only 4 of 14 grade III/IV GVHD sibling pairs were identical for CD31 genotype, compared to 25 of 32 grade 0 patients. The much larger size of our data set suggests that the results of Behar et al may represent a chance occurrence. Stratification of the Behar et al data by HLA A2 genotype revealed that the observed effect of CD31 mismatch was restricted to the group of patients negative for the HLA A2 allele (0 of 8 grade III/IV patients were identical for CD31). Three of these 8 "nonidentical" pairs were actually compatible. This contrasts with the 117 HLA A2 negative grade III/IV pairs in our study, for which 68% were CD31 identical. Based on these results, we concluded that the codon 125 polymorphism at CD31 is unlikely to represent a significant risk factor for GVHD in human HLA identical sibling BMT.[13] Our observations do not support the use of CD31 genotype analysis in selection of BMT marrow donors.

Looking to the Future

The critical assumption for the success of a genomics-based approach outlined above is the existence of only a limited number of minor *H* genes that are clinically significant for determining GVHD (i.e., "major minors"). If a large complex panel of minor *H* genes interact in different ways in each GVHD patient, resulting in an essentially unique set of mismatches in each sibling donor pair, then this genetic approach will have a low probability of success. Alternatively, if there are a limited number of immunodominant minor *H* genes, the chance of success should be high. A number of other complex genetic diseases are currently being approached in this way, including asthma, type I and type II diabetes, and hypertension. Several of these studies have already successfully identified significant genetic contributions from limited numbers of loci. Of note, type I diabetes, one of the first reported successes for the genome-wide linkage approach,[14] is only partially determined by genetic factors, with ~20-40% concordance among identical twins. In contrast, as described above, GVHD can be considered as being 100% heritable.

At present, the total number and distribution of *GVHD* genes is unknown and a subject of considerable discussion and speculation. The completion of a detailed genome scan as

Table 11.2. CD31 genotyping of patients and donors

	GVHD Status	
	Grade 0 (N=74)	Grade III/IV (N=227)
Genotypes of donor and recipient		
Identical	51	143
Nonidentical		
Both homozygous	2	10
Heterozygous donor, homozygous recipient	11	32
Homozygous donor, heterozygous recipient	10	42
Genotypes matched for identity		
Identical	51 (0.69)	143 (0.63) p^*=0.36
Nonidentical	23 (0.31)	84 (0.37)
	p=0.010	p=0.27
Genotypes matched for compatibility		
Compatible	63 (0.85)	175 (0.77) p=0.14
Incompatible	11 (0.15)	52 (0.23)
	p=0.14	p=0.71

* p values at the far right represent comparisons of Grade III/IV GVHD sib pairs to Grade 0 GVHD sib pairs; p values below subgroups represent comparisons of the observed to the expected.[Taken from ref. 13, with permission.]

described here should provide the first objective data to allow assessment of the genetic complexity of the minor *H* genes clinically significant for GVHD. In addition, the DNA samples assembled for this patient cohort provide a powerful tool for testing the importance of future candidate genes identified by other approaches.

Finally, recent advances in human genome research should provide dramatically increased power for this type of linkage analysis. Specifically, a large number of single nucleotide polymorphisms (SNPs) is being developed as a component of the human genome project. It is estimated that within 2 years a panel of approximately 20,000 such SNPs may become available.[15] Genotyping may also become considerably less expensive and more rapid using DNA "chip" technology. These technical advances should significantly increase our power to detect more subtle contributions of minor *H* genes and allow us to dissect the complex genetic interactions underlying GVHD.

References

1. Ferrara JLM, Deeg HJ. Graft-versus-host disease. N Engl J Med 1991; 324:667.
2. Vogelsang GB, Hess AD. Graft-versus-host disease: New directions for a persistent problem. Blood 1994; 84:2061.
3. Armitage JO. Bone marrow transplantation. N Engl J Med 1994; 330:827.
4. Lindahl KF. Minor histocompatibility antigens. Trends Genet 1991; 7:219.
5. van Els CACM, D'Amaro J, Pool J et al. Immunogenetics of human minor histocompatibility antigens: Their polymorphism and immunodominance. Immunogenetics 1992; 35:161.
6. Beatty PG. National Heart, Lung, and Blood Institute (NHLBI) workshop on the importance of minor histocompatibility antigens in marrow transplantation. Exp Hematol 1997; 25:548.

7. Meadow L, Wang W, den Haan JM et al. The HLA-A*0201-restricted H-Y antigen contains a posttranslationally modified cysteine that significantly affects T-cell recognition. Immunity 1997; 6:273-281.
8. Behar E, Chao NJ, Hiraki DD et al. Polymorphism of adhesion molecule CD31 and its role in acute graft-versus-host disease. N Engl J Med 1996; 334:286.
9. Den Haan JMM, Sherman NE, Blokland E et al. Identification of a graft-versus-host disease-associated human minor histocompatibility antigen. Science 1995; 268:1476.
10. Goulmy E, Schipper R, Pool J et al. Mismatches of minor histocompatibility antigens between HLA-identical donors and recipients and the development of graft-versus-host disease after bone marrow transplantation. N Engl J Med 1996; 334:281.
11. Den Haan JMM, Meadows LM, Wang W et al. The minor histocompatibility antigen HA-1: A diallelic gene with a single amino acid polymorphism. Science 1998; 279:1054.
12. Gubarev MI, Jenkin JC, Leppert MF et al. Localization to chromosome 22 of a gene encoding a human minor histocompatibility antigen. J Immunol 1996; 157:5448.
13. Nichols WC, Antin JH, Lunetta KL et al. Polymorphism of adhesion molecule CD31 is not a significant risk factor for graft-versus-host disease. Blood 1996; 88:4429.
14. Davies JL, Kawaguchi Y, Bennett ST et al. A genome-wide search for human type 1 diabetes susceptibility genes. Nature 1994; 371:130.
15. Kruglyak L. The use of a genetic map of biallelic markers in linkage studies. Nat Genet 1997; 17:21.

CHAPTER 12

Mouse Models
for Graft-Versus-Host Disease

Robert Korngold, Marc A. Berger, Debbie Statton and Thea M. Friedman

Bone Marrow Transplantation

Allogeneic bone marrow transplantation (BMT) has developed into an important therapeutic approach for the treatment of hematologic disorders, and in particular several forms of leukemia.[1] However, BMT is still hampered by the development of donor T cell-mediated graft-versus-host disease (GVHD).[2-5] Although the approach of depleting T cells from the donor marrow inoculum has been successful in minimizing the incidence of GVHD, most clinical studies have also reported increased incidence of graft failure, opportunistic infection, and leukemic relapse.[6] In this respect, it would be highly advantageous to be able to manipulate the donor marrow or subsequent T-cell infusions in such a way as to selectively deplete alloreactive T cells responsible for GVHD, while retaining T cells that could support engraftment and/or decrease the occurrence of infection and leukemic relapse.

Minor Histocompatibility Antigens and GVHD

Multiple minor histocompatibility (H) antigens are involved in the induction of GVHD following major histocompatibility complex (MHC)-matched allogeneic BMT.[7-11] Minor H antigens can be presented by both MHC class I and class II molecules,[12,13] and are recognized by CD8[+] cytotoxic T lymphocytes (CTL) and CD4[+] helper-type cells, respectively. Experimentally, it is clear that both CD8[+] and CD4[+] T-cell subsets may generate lethal GVHD to minor H antigens independently or through a cooperative pathogenic mechanism.[14-18] It is also apparent that either identical or completely independent genetic loci may give rise to the actual minor H antigen peptides presented by MHC class I and II molecules, and that numerous such loci may exist between MHC-matched unrelated individuals.[19-21] This notion raises a central issue of whether the GVHD T-cell response to allogeneic minor H antigens is so vast that it would never allow specific manipulation, or that it is oligoclonal and limited to a few immunodominant minor H antigens that are responsible for disease.

Different Immunodominant Minor H Antigens Detected In Vitro and in GVHD

The concept of immunodominance of minor H antigens was by generation of CTL from C57BL/6By (B6) mice against BALB.B minor H antigens. Despite an estimated antigenic difference of at least 29 minor H antigens between the genetic backgrounds of these two strains,[22] it was found that the in vitro CTL responses were directed to only a few immunodominant antigens.[23] These minor H antigens were detected because of their differential expression in a panel of target cells from the CXBE, G, I, J and K recombinant inbred (RI) strains originally generated by crosses of F_2 generations from a B6 and BALB/c mating.[24] Each of these CXB RI strains contain a different distribution of B6 and BALB

Minor Histocompatibility Antigens: From the Laboratory to the Clinic, edited by Derry Roopenian and Elizabeth Simpson. ©2000 Landes Bioscience.

alleles. Since they are all $H2^b$-matched, B6 T-cell responses to antigen presenting cell stimulation from the CXB RI are directed against whatever BALB minor H antigens are expressed. Furthermore, B6 responses against the BALB.B strain would potentially encompass all of the responses to minor H antigens that can be found to the CXB RI strains. However, GVHD studies involving the transplantation of B6 T cells into irradiated CXB RI strains indicated that the immunodominant minor H antigens in vitro hierarchy did not correlate with GVHD potential.[25] GVHD was not evident in the CXBG and CXBK strains, which expressed first-order immunodominant minor H antigens for CTL activity, to which most of the in vitro responses were directed. In contrast, strong GVHD responses were observed in CXBE strain recipients, which expressed only second-order immunodominant minor H antigens (they could only generate CTL responses in the absence of the first-order immunodominant minor H antigens).

A recent inter-strain GVHD response analysis with the CXB RI strains suggested that a minimum of two distinct MHC class I-restricted minor H antigens could account for disease in the B6->BALB.B combination.[26] One immunodominant minor H antigen was shared by CXBE, CXBI, and CXBJ strains and the second minor H antigen was uniquely expressed by the BALB.B strain. Consistent with these observations, all positive B6 GVHD responses in the BALB.B and CXB RI strains involved mediation by $CD8^+$ T cells, most of which appeared to be dependent upon $CD4^+$ T-cell help.[18] In addition, $CD4^+$ T cells independently caused a high level of lethal GVHD in BALB.B, and to a lesser extent in CXBI, recipients.

Potential Mechanisms of Immunodominance

The phenomenon of immunodominance amongst minor H antigens may reflect their comparative abilities to compete successfully for antigen-binding of the appropriate MHC molecules on APC. Alternatively, the affinity/avidity of the interaction of the specific TCR for minor H antigen peptide/self-MHC may also be an important factor of minor H antigen immunodominance. In addition, T cells that respond more vigorously to strong minor H antigens may downregulate weaker responses to other minor H antigens via cytokine production. Each of these potential mechanisms or combinations thereof could account for immunodominance at the level of T-cell recognition and responsiveness. However, potential differences between the in vitro and in vivo stimulating microenvironments and other variables related to the particular alloreactive combination may have a bearing on T-cell activation and consequently which minor H antigens appear immunodominant. Recent studies comparing CTL immunodominant specificities and skin graft rejection in the B6 anti-BALB.B and CXB RI strain combinations have found a distinct lack of correlation between the two responses, with CTL recognizing only a limited number of minor H antigens operative in vivo.[27]

A similar lack of correlation with CTL immunodominant specificities had previously been found in GVHD.[18,25] For GVHD then, the relevant question is how extensive are the number of minor H antigens that are actually involved in GVHD development? There must be certain limitations and qualifications for minor H antigens to serve in this capacity. At least two of these criteria are immunogenicity and tissue distribution. In terms of the latter, it is known that some minor H antigens have unique tissue expression[28,29] and it would be expected that for optimum GVHD pathogenesis, antigen should be available not only in cells of hematopoietic origin, but in the primary target organs, including the gut, liver, and skin. The lack of GVHD in irradiation chimeric models which only have minor H antigens expressed in the host hematopoietic compartment supports the notion that the presence of antigen in both locations is required.[30]

The questions of immunodominant minor H antigens and the heterogeneity of GVHD T-cell responses remain important questions to be addressed before a full understanding of the disease process can be achieved. In the following sections, we will summarize our most recent studies of both the CD8$^+$ and CD4$^+$ T-cell repertoires involved in the B6->BALB.B GVHD response, as initial steps toward gaining this level of understanding.

Analysis of T-Cell Receptor Gene Usage During GVHD

Thoracic Duct Lymphocyte Analysis of Donor Anti-Host Minor H Antigen Responses

To gain insight into the question of the nature and extent of the CD8$^+$ T-cell GVHD response to minor H antigens in the B6->BALB.B combination, we took the approach of analyzing positively-selected minor H antigens-specific T cells collected from the thoracic duct lymphocyte (TDL) pool of recipient mice. Since recently expanded and activated T cells enter the thoracic duct after initially encountering minor H antigens in the spleen and lymph nodes, the TDL pool is an excellent site for monitoring T-cell responses. Particularly important, is that the cells can be retrieved with high viability ($\geq$99%) in contrast to retrieval of cells from the peripheral lymphoid organs, and if properly stimulated TDL will contain significant levels of blast-like T cells.[31] Due to the overall low frequency responses involved in anti-minor H antigen reactivity and the slower development of primary CD8$^+$ T-cell responses in vivo, phenotypic skewing of Vβ families was not detectable over a period of 3-8 days post-transplantation of naive B6 T cells, although significant skewing occurred in the CD4$^+$ T-cell population.[32] The cannulation of mice at later time points was impractical, since the mice began developing GVHD related symptoms and could not survive the procedure. As an alternative, anti-host minor H antigen T cells were first expanded by presensitization and boosting of donor B6 mice in vivo with host-type splenocytes. We then investigated the response patterns of these donor T cells (1-1.5 x 10^7) when placed in a GVHD-inducing environment in the irradiated (825cGy) BALB.B or CXBE recipients. The thoracic ducts of recipients were cannulated 5 days later, based on previously well-established procedures,[32,33] and TDL collected over a 20 hour period (Fig. 12.1). The positively-selected TDL were greater than 95% CD3$^+$ and consisted of 55-67% CD4$^+$ and 33-45% CD8$^+$ T cells. In addition, a significant percentage (20-25%) of the TDL cells were blast-like in size, in contrast to TDL collected from B6 mice transplanted with T cells (1.5 x 10^7) from syngeneic-presensitized B6 mice which yielded few blast-like cells and were >98% CD4$^+$. Therefore, the TCR Vβ cell repertoire analyses of TDL CD8$^+$ T cells from GVHD recipients was compared to the normal TCR Vβ cell repertoire of CD8$^+$ T cells from normal B6 spleens.

Donor CD8$^+$ T-Cell Responses Have Limited Vβ Usage

TCR Vβ cell repertoire analysis of the TDL was initially conducted by two-color flow cytometry using a panel of available Vβ-specific mAb along with a CD8-specific mAb. Of the 13 Vβ families assessed, the B6->BALB.B TDL failed to exhibit any significant expansion of a Vβ family, whereas B6->CXBE TDL exhibited significant expansions of the Vβ10 and Vβ14 families, relative to naive B6 splenic CD8$^+$ T cells.[34] In the B6->CXBE TDL, the expansions were from 6.1% to 9.9% and 3.3% to 6.8% of the total Vβ population for the Vβ10 and 14 families, respectively. The observation of expansions of a limited number of Vβ families in the B6->CXBE TDL suggested the involvement of a limited in vivo CD8$^+$ T-cell immune response to multiple minor H antigen differences involved in GVHD.

In order to ensure that the TDL cells were capable of inducing GVHD, positively-selected T cells from B6->BALB.B TDL (1.4 x 10^7) were transplanted along with 2 x 10^6 anti-Thy1.2 and complement-treated bone marrow cells (ATBM) into lethally irradiated (825cGy) BALB.B or CXBE recipients. By day 25, both the BALB.B and CXBE recipients exhibited the clinical symptoms of GVHD, including diarrhea, ruffled fur and weight loss (>20%),

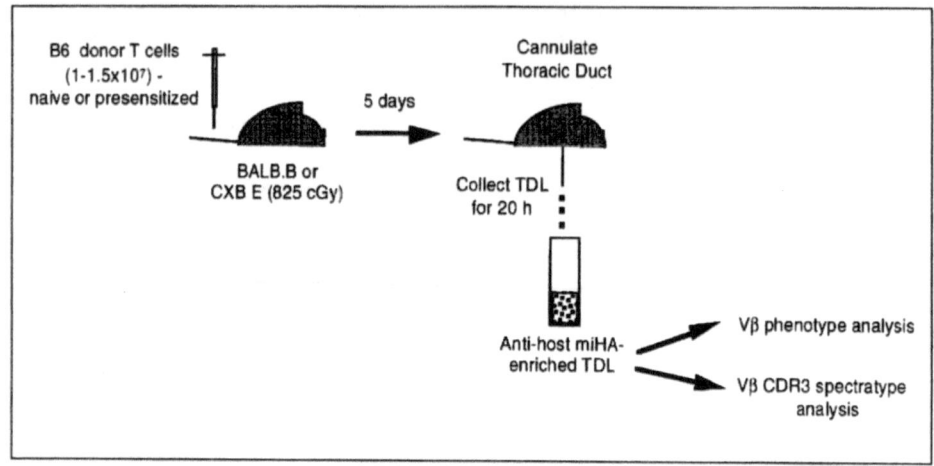

Fig. 12.1. TDL collection procedure.

compared to control recipients of donor ATBM alone. All of the BALB.B recipients of TDL died by day 50 [median survival time (MST) of 30 days] and all of the CXBE recipients died by day 72 (MST of 50 days).

TCR Vβ CDR3 Spectratype Analysis

The CDR3 sequence encoding the variable β chain V-D-J junctional region defines the unique TCR clonotype/s specific for an antigen, and acts as a fingerprint for the T-cell lineage(s) bearing it. PCR-based analysis, known as CDR3-size spectratyping was originally developed by Gorski et al[35,36] for analysis of alloreactive T cells following bone marrow transplantation, but the technique has more recently been applied for repertoire analysis in other disease states.[37-39] For our own studies involving anti-minor H antigen responses, spectratyping was used to precisely determine the nature of the Vβ family repertoire in the B6→ BALB.B and B6→ CXBE CD8⁺ TDL, as a reflection of the initial GVHD response. The band distribution of a normal spectratype follows a Gaussian distribution with the center bands of median CDR3 length being most intense and the outer bands of longer and shorter lengths being less intense (Fig. 12.2). In contrast, the band distribution pattern of a skewed spectratype does not exhibit a Gaussian distribution, but has some bands more intense relative to the control distribution (Fig. 12.2). The intensity of these bands can be readily quantitated by either densitometric scanning of autoradiographs or fluoroimaging analysis. The spectratyping approach also has a tremendous advantage over phenotypic analysis of the T-cell repertoire in the potential ability to isolate unique TCR clonotypes and define the reactive TCR sequences involved.

The CDR3-size spectratypes of seventeen Vβ families expressed in B6 mice (Vβ1-16 and Vβ20) were examined for band skewing suggestive of an oligoclonal expansion of particular T-cell lineages.[34] The results, summarized in Table 12.1, supported the findings of the phenotypic analysis: Vβ6 exhibited biased CDR3 usage in the B6→BALB.B CD8⁺ TDL and Vβ10 and Vβ14 were skewed in the B6→CXBE CD8⁺ TDL. In addition, the spectratype analysis revealed that both strain combinations exhibited skewing of bands in the Vβ1, 6, 8, 9, 10, and 14 families. In all cases, the same CDR3 size band was enhanced in the same Vβ family in the two combinations. However, in the V14 family, additional bands were enhanced in the B6→BALB.B CD8⁺ TDL spectratype which were not skewed in the

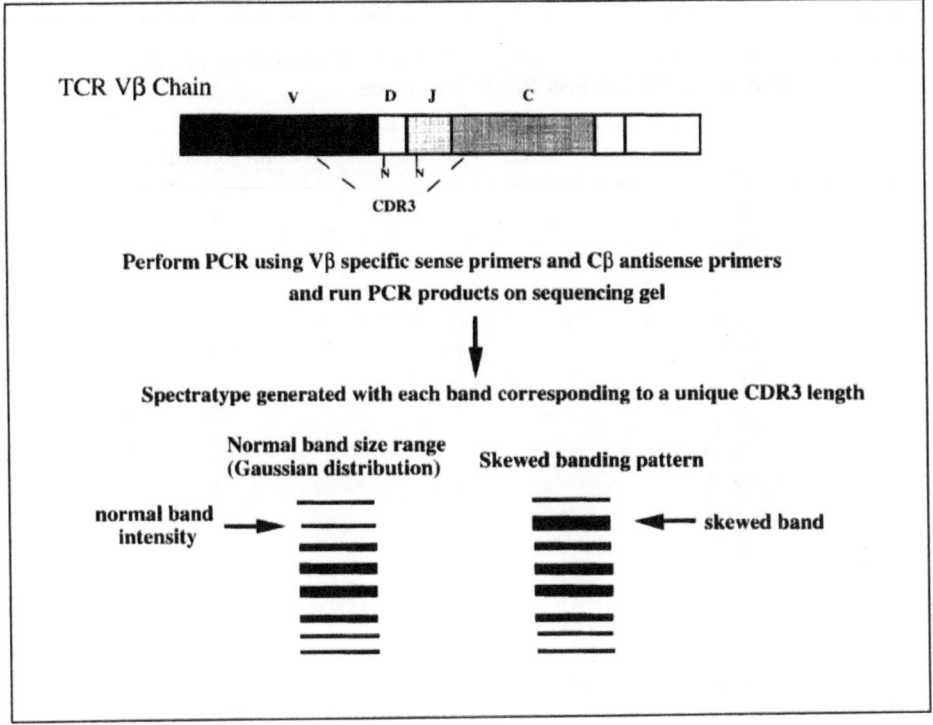

Fig. 12.2. CDR3 spectratyping approach.

B6→CXBE CD8+ response. Interestingly, biased CDR3 usage in the Vβ4 family was unique to the B6→BALB.B CD8+ TDL and suggested a response to a BALB.B-specific minor H antigens. Overall, the observation of limited heterogeneity in the CD8+ T-cell response in GVHD is consistent with the involvement of a limited number of immunodominant minor H antigens in GVHD, as has been previously proposed.[40] Similar observations have also been made at the level of local liver GVHD pathogenesis in the B10.D2→BALB/c strain combination.[41]

There are several possible factors responsible for the level of limited heterogeneity observed by CDR3-size spectratyping analysis. GVHD may indeed be mediated by a heterogenous population of alloreactive CD8+ T cells that recognize several minor H antigens or multiple epitopes of the same Ag in the host. It is also possible that multiple TCRs can recognize the same immunodominant Ags, since T-cell recognition of Ag involves the interaction of both Vβ and Vα chains.[42] Each of the responding Vβ chains could associate with a different Vα chain to recognize the same epitope. This type of phenomenon has been observed in the diversity of CD4+ T-cell responses to influenza antigenic determinants.[43]

In an attempt to confirm the involvement of individual Vβ family constituents in lethal GVHD responses, as suggested by the spectratype analysis, cells were positively selected by magnetic cell sorting and tested for GVHD capacity. CD8+ Vβ14+ T cells were isolated from B6 mice presensitized to CXBE minor H antigens and then transplanted into irradiated (825 cGy) BALB.B recipients (1x10⁵ >99%-enriched T cells) along with 3 x 10⁶ naive B6 CD4+ T cells (to provide T helper cell function) and 2x10⁶ ATBM cells. BALB.B recipients of either 4 x 10⁶ unseparated CXBE-presensitized B6 CD8+T cells or isolated Vβ14+ CD8+ T cells exhibited the clinical symptoms of GVHD between days 45-50, including weight loss

Table 12.1. Summary of CD8⁺ TCR Vβ CDR3-biased utilization of TDL from B6 αBALB.B and B6 αCXBE mice

	B6 → BALB.B	B6 → CXBE
Vβ 1	+	+
Vβ 2	-	-
Vβ 3	-	-
Vβ 4	+	-
Vβ 5.1,2	-	-
Vβ 6	+	+
Vβ 7	-	-
Vβ 8.1,2,3	+	+
Vβ 9	+	+
Vβ 10	+	+
Vβ 11	±	±
Vβ 12	-	-
Vβ 13	-	-
Vβ 14	+	+
Vβ 15	-	-
Vβ 16	-	-
Vβ 20	-	-

± definitive bias could not be determined

(22% and 30%, respectively). By day 76, 60% (MST of 49 days) of the BALB.B mice that received the unseparated CD8⁺ T cells and 80% (MST of 67 days) of those given the Vβ14⁺ CD8⁺ T cells died, compared to 20% mortality in the ATBM control group. In a similar manner, lethally irradiated CXBE recipients of either unseparated (3 x 10⁷ CXBE presensitized) or Vβ14⁺ (3.5 x 10⁶ >84%-enriched) T cells exhibited the clinical symptoms of GVHD by day 40, including weight loss (17% and 25%, respectively). By day 45, mice that received injections of Vβ14⁺ enriched T cells had 80% fatality (MST of 48 days), and by day 70, the mice that had been injected with unseparated T cells had 40% fatality, as compared to 100% survival in the control ATBM group. Overall, these experiments suggest that Vβ14⁺ T cells can mediate lethal GVHD. Further analysis of these responses are necessary to establish whether a single or limited number of TCR specificities are involved from this family and the distribution of target sites for disease.

Overall, the CD8⁺ T-cell repertoire results are consistent with a limited number of minor H antigens that fit the criteria for GVHD. In this respect, our recent study of the CXB interstrain lethal GVHD responses suggested the minimal involvement of two distinct immunodominant minor H antigens in the B6→BALB.B GVHD response.[26] One class I restricted minor H antigens (GVH-1) appeared to be shared by the CXBE strain, while the second antigen (GVH-2) was uniquely expressed by the BALB.B strain. The biased CDR3-size skewing of Vβ1, 6, 8, 9, 10, and 14 families in both the BALB.B and CXBE recipients may represent a response to common minor H antigens shared by BALB.B and CXBE mice, i.e., GVH-1. Although the same CDR3-size bands were skewed in each corresponding Vβ family (except for Vβ14 in which additional bands were skewed in the BALB.B recipients), final identity of these TCRs will depend upon sequence analysis of the CDR3 segments involved in each response. Furthermore, the biased CDR3 usage of Vβ4 in the BALB.B recipients, not present in the spectratype of the CXBE recipients, might represent a response to the unique BALB.B minor H antigen, i.e., GVH-2. Experiments are underway to determine if positively-selected CD8$^+$ Vβ4$^+$ T cells will induce GVHD in BALB.B but not CXBE recipients.

Donor CD4$^+$ T-Cell Responses Have Limited Vβ Usage

As mentioned earlier, CD4$^+$ T cells can certainly participate in a GVHD response to minor H antigens by providing helper function for the generation of minor H antigen-specific CTL.[15,18] Less frequently, CD4$^+$ T cells alone have been capable of inducing lethal GVHD directed to minor H antigens, although the factors that control these responses are not well understood.[14,15,17] To investigate the complexity and diversity of the minor H antigen-specific CD4$^+$ T-cell response in GVHD, we have again focused our attention upon the C57BL/6 (B6->BALB.B (825 cGy) strain combination, particularly because it is a model in which CD4$^+$ T cell-mediated lethal GVHD does develop.[18] Similar to that performed for the CD8$^+$ T-cell component, phenotypic analysis of the TCR Vβ cell repertoire of CD4$^+$ T cells collected in the TDL of irradiated (825 cGy) recipient mice during the initial development of GVHD (days 4-6) suggested that there was also a limited T cell response to minor H antigens.[32] Flow cytometry indicated notable expansions of the CD4$^+$ Vβ3, 6, 7, 8, and 9 families. By comparison, in the B6->CXBE recombinant inbred strain combination, in which CD4$^+$ T cells do not cause lethal GVHD and likely involves recognition of only some of the minor H antigens that are expressed by BALB.B mice, the repertoire analysis indicated expansions of only the Vβ3, 7 and 9 families. The Vβ3 expansion was attributed to the known response to the MTV-6 superantigen expressed in both BALB.B and CXBE mice.[44,45] This Vβ3 response was extensive by day 4 post-transplantation with an expansion from the normal 2% to 20% of the TDL population, but decreased dramatically afterwards. It is not clear what the role, if any, of this superantigen response in GVHD is, although it in itself apparently does not cause lethal GVHD, since CXBE mice also have the CD4$^+$ T-cell response but fail to develop pathological effects.

CDR3-size spectratyping was also performed to obtain a more precise characterization of the CD4$^+$ anti- minor H antigen response. The results, summarized in Table 12.2, demonstrated that there was overlapping biased expansion in Vβ2, 6, 7, 8, 9, 10, and 14 families in both the B6$\rightarrow$BALB.B and B6$\rightarrow$CXBE CD4$^+$ TDL populations. The same CDR3-size band was enhanced in each of the skewed Vβ families between the two strain combinations. In the case of Vβ3, biased CDR3 utilization was not observed since a polyclonal response was consistent with that expected from a superantigen stimulation. Overall, the spectratyping results were still suggestive of limited oligoclonal anti-minor H antigen CD4$^+$ T-cell responses, and the approach proved to be clearly more sensitive than phenotypic analysis in detecting the full scope of the in vivo CD4$^+$ T-cell response.

In order to determine if there would be any changes in the Vβ family repertoires upon continued exposure to minor H antigens, allowing for potential expansion of low frequency minor H antigen-specific B6 CD4$^+$ T cells, B6$\rightarrow$BALB.B and B6$\rightarrow$CXBE TDL were analyzed by CDR3-size spectratyping 5 days after transplantation with T cells from host-presensitized B6 donor mice. In this case, the B6$\rightarrow$BALB.B and B6$\rightarrow$CXBE CD4$^+$ TDL again exhibited the expected overlapping biased CDR3 usage in the Vβ2, 6, 7, 8, 9, 10, and 14 families (Table 12.2). Once again, the same CDR3 size band was enhanced in each skewed Vβ family between the two groups. In addition, the Vβ 13 families appeared to be skewed in both strains of mice. We are beginning to analyze sequential histological sections from B6 >BALB.B mice undergoing GVHD for infiltrating T-cell Vβ expression by either immunocytochemistry or in situ hybridization techniques. In this way, we hope to discern whether the spectratype results correlate with GVHD-mediating T cells.

Table 12.2. Summary of CD4⁺ TCR Vβ CDR3-biased utilization of TDL from B6 → BALB.B and B6 → CXBE mice

	Naive B6 → BALB.B	Naive B6 → CXBE	Presensitized B6 → BALB.B	Presensitized B6 →CXBE
Vβ1	±	±	±	±
Vβ2	+	+	+	+
Vβ3	-	-	-	-
Vβ4	-	-	-	-
Vβ5.1,2	-	-	-	-
Vβ6	+	+	+	+
Vβ7	+	+	+	+
Vβ8.1,2,3	+	+	+	+
Vβ9	+	+	+	+
Vβ10	+	+	+	+
Vβ11	±	±	±	±
Vβ12	-	-	-	-
Vβ13	±	±	+	+
Vβ14	+	+	+	+
Vβ15	-	-	-	-
Vβ16	-	-	-	-

± definitive bias could not be determined

CDR3 Sequence Analysis in Reactive CD4⁺ T-Cell Vβ Families

Although the same CDR3-size bands were skewed in most corresponding Vβ families, final identity of these TCRs will depend upon sequence analysis of the CDR3 segments involved in each response. Although one cannot directly deduce the minor H antigen epitope sequence from an analysis of the TCR CDR3 sequence involved in its recognition, it can help to clarify the extent of clonality of the responses within each Vβ family, and perhaps even reveal similarities between the responses in different Vβ families. In this respect, we have begun to analyze the CD4⁺Vβ6⁺ B6→BALB.B TDL response by sequence cloning. cDNA was purified from the skewed CDR3-size band of the Vβ6 spectratyping gel. After amplification by PCR, the cDNA was cloned, yielding 113 clones, and sequenced by standard dideoxy chain termination reaction chemistry and an automated sequencing system. The results indicated that cloned Vβ6 TCR was heavily biased in the usage of Jβ1.3 and Jβ2.4 segments, compared to cloned Vβ6 TCR from naive B6 CD4⁺ T cells. It has previously been demonstrated that Jβ usage in normal B6 mice is non-random with preferential usage of the Jβ2 cluster.[46] There were no Vβ6Jβ1.3 rearrangements detected in naive CD4⁺ TCR compared to 10.6% Jβ1.3 usage by B6→BALB.B Vβ6 TCR. There was also heavy biased usage in the Jβ2.4 element (71.7% versus 25.0% in naive TCR). Overall, the Vβ6 TCR from naive B6 CD4⁺ T cells exhibited a much more heterogeneous usage of Jβ elements in comparison with the B6→BALB.B CD4⁺ TDL. Thus, it appears that in the Vβ6 family there is a limited oligoclonal minor H antigen-driven repertoire skewing in the early phase of GVHD development.

Similar to that discussed above for CD8⁺ T cells, the observed heterogeneity by CDR3-size spectratyping analysis of the naive and presensitized CD4⁺ TDL response may again indicate overlapping recognition of only a few immunodominant minor H antigens

antigens. Again, there may be multiple TCR Vβ family involvement, since T-cell recognition of an antigen likely involves the interaction of both Vβ and Vα chains.[39] Therefore, the overall avidity of TCR/minor H peptide/MHC interaction may be equivalent between TCRs of different Vβ families with differences in CDR3 binding sites being compensated for by Vα engagement. Alternatively, there may be several minor H antigen epitopes involved in the GVHD response. In this respect, Wettstein has recently reported a minimum of six different CXB RI strain related class II-restricted minor H antigens encoded by independently segregating genes that can stimulate in vitro B6 CD4[+] T cells.[47] Whether these same antigens are operative in vivo in the GVHD pathogenesis remains to be determined.

Conclusion

Characterization of the TCR Vβ cell repertoire of minor H antigen-specific T cells mediating GVHD will help us to understand the scope of these complex responses and the nature of the antigens responsible for their induction. The results, thus far, indicate that there is a limited heterogenous response in both the donor CD8[+] and CD4[+] T-cell populations during the early development of lethal GVHD in the B6→ BALB.B strain combination. This observation in turn is suggestive of a limited number of immunodominant minor H antigens that fit the requirements for generation of disease in vivo. In addition, the overlapping spectratype data from both the B6→ BALB.B and B6→CXBE strain combinations would argue against the competition between these immunodominant minor H antigens for recognition in the GVHD setting; i.e., if an immunodominant minor H antigen can be expressed by host tissue, the donor T cells will respond despite the presence of other immunodominant epitopes. This may mean that there are no hidden minor H antigen hierarchies that can cause GVHD. If these observations can be extrapolated to humans, it would be useful to define immunodominant minor H antigens associated with HLA alleles that are present at high frequencies in the population. Indeed, this arduous task has already been initiated by Goulmy et al.[48] Ultimately, with the future development of diagnostic capabilities for these immunodominant minor H antigens, it may be possible for the application of new targeted strategies for prevention of GVHD.

References

1. Thomas ED, Clift RA, Storb R. Indications for marrow transplantation. Ann Rev Med 1984; 35:1-9.
2. Gale RP. Graft-versus-host disease. Immunol Rev 1985; 88:193-214.
3. Santos GW, Hess AD, Vogelsang GB. Graft-versus-host reactions and disease. Immunol Rev 1985; 88:169-192.
4. Deeg HJ, Storb R. Graft-versus-host disease: Pathophysiological and clinical aspects. Ann Rev Med 1984; 35:11-24.
5. Ferrara J, Deeg HJ. Graft-versus-host disease. New Engl J Med 1991; 324:667-674.
6. Kernan NA. T-cell depletion for prevention of graft-versus-host disease. In: Forman SJ, Blume KG, Thomas ED, eds. Bone marrow transplantation. Cambridge: Blackwell Scientific Publications, 1994:124-135.
7. Korngold R, Sprent J. Lethal graft-versus-host disease following bone marrow transplantation across minor histocompatibility barriers in mice. Prevention by removing mature T cells from marrow. J Exp Med 1978; 148:1687-1698.
8. Mathe G, Pritchard LL, Halle-Pannenko O. Mismatching for minor histocompatibility santigens in bone marrow transplantation: Consequences for the development and control of severe graft-versus-host disease. Transplant Proc 1979; 11:235-239.
9. Perreault C, Decary F, Brochu S et al. Minor histocompatibility antigens. Blood 1990; 76:1269-1280.
10. Goulmy E, Voogt P, van Els C et al. The role of minor histocompatibility antigens in GVHD and rejection: A mini-review. Bone Marrow Transplant 1991; 1:49-51.

11. Martin PJ. Increased disparity for minor histocompatibility antigens as a potential cause of increased GVHD risk in marrow transplantation from unrelated donors compared with related donors. Bone Marrow Transplant 1991; 8:217-223.

12. Bevan MJ. The major histocompatibility complex determines susceptibility to cytotoxic T cells directed against minor histocompatibility antigens. J Exp Med 1975; 142:1349-1364.

13. Hurme M, Hetherington CM, Chandler PR et al. Cytotoxic T-cell responses to HY: mapping of the *Ir* genes. J Exp Med 1978; 147:758-767.

14. Korngold R, Sprent J. Variable capacity of L3T4[+] T cells to cause lethal graft-versus-host disease across minor histocompatibility barriers in mice. J Exp Med 1987; 165:1552-1564.

15. Hamilton BL. L3T4-positive T cells participate in the induction of graft-vs-host disease in response to minor histocompatibility antigens. J Immunol 1987; 139:2511-2515.

16. Korngold R. Lethal graft-versus-host disease in mice directed to multiple minor histocompatibility antigens: Features of CD8[+] and CD4[+] T-cell responses. Bone Marrow Transplant 1992; 9:355-364.

17. Miconnet I, Huchet R, Bonardelle D et al. Graft-versus-host mortality induced by noncytolytic CD4[+] T-cell clones specific for non-H2 antigens. J Immunol 1990; 145:2123-2131.

18. Berger M, Wettstein PJ, Korngold R. T-cell subsets involved in lethal graft-versus-host disease directed to immunodominant minor histocompatibility antigens. Transplant 1994; 57:1095-1102.

19. Davis AP, Roopenian DC. Complexity at the mouse minor histocompatibility locus H4. Immunogen 1990; 31:7-12.

20. Roopenian DC, Davis AP, Christianson GJ et al. The functional basis of minor histocompatibility loci. J Immunol 1993; 151:4595-4605.

21. Roopenian DC. What are minor histocompatibility loci? A new look at an old question. Immunol Today 1992; 13:7-10.

22. Bailey DW, Mobraaten LE. Estimates of the number of loci contributing to the histoincompatibility between C57BL-6 and BALB-c strains of mice. Transplant 1969; 7:394-400.

23. Wettstein PJ. Immunodominance in the T-cell response to multiple non-H-2 histocompatibility antigens. II. Observation of a hierarchy among dominant antigens. Immunogenetics 1986; 24:24-31.

24. Bailey DW. Recombinant-inbred strains. An aid to finding identity, linkage, and function of histocompatibility and other genes. Transplant 1971; 11:325-327.

25. Korngold R, Wettstein PJ. Immunodominance in the graft-vs-host disease T-cell response to minor histocompatibility antigens. J Immunol 1990; 145:4079-4088.

26. Korngold R, Leighton C, Mobraaten LE et al. Inter-strain graft-versus-host disease T-cell responses to immunodominant minor histocompatibility antigens. Biol Blood Marrow Transplant 1997; 3:57-64.

27. Nevala WK, Paul C, Wettstein PJ. Reduced diversity of CTLs specific for multiple minor histocompatibility antigens relative to allograft rejection in vivo. J Immunol 1997; 158:1102-1107.

28. Johnson LL, Bailey DW, Mobraaten LE. Genetics of histocompatibility in mice. IV. Detection of certain minor (non-H-2) H antigens in selected organs by the popliteal node test. Immunogenetics 1981; 14:63-71.

29. Griem P, Wallny H-J, Falk K et al. Uneven tissue distribution of minor histocompatibility protein versus peptides is caused by MHC expression. Cell 1991; 65:633-640.

30. Korngold R, Sprent J. Features of T cells causing H-2-restricted lethal graft-vs.-host disease across minor histocompatibility barriers. J Exp Med 1982; 155:872-883.

31. Sprent J. Role of the H-2 complex in induction of T helper cells in vivo. I. Antigen-specific selection of donor T cells to sheep erythrocytes in irradiated mice dependent upon sharing of H-2 determinants between donor and host. J Exp Med 1978; 148:478-489.

32. Berger MA, Korngold R. Immunodominant CD4[+] T-cell receptor Vβ cell repertoire involved in graft-versus-host disease responses to minor histocompatibility antigens. J Immunol 1997; 159:77-85.

33. Korngold R, Bennink JR. Collection of mouse thoracic duct lymphocytes. Meth Enzymol 1984; 108:270-274.
34. Friedman TM, Gilbert M, Briggs C et al. Repertoire analysis of CD8[+] T-cell responses to minor histocompatibility antigens involved in graft-versus-host disease. J Immunol 1998; 161:4-8.
35. Gorski J, Yassai M, Zhu X et al. Circulating T-cell repertoire complexity in normal individuals and bone marrow recipients analyzed by CDR3 size spectratyping. J Immunol 1994; 152:5109-5119.
36. Gorski J, Piatek T, Yassai M et al. Improvements in repertoire analysis by CDR3 size spectratyping bifamily PCR. Ann NY Acad Sci 1995; 76:99-102.
37. Kolowos W, Herrmann M, Ponner BB et al. Detection of restricted junctional diversity of peripheral T cells in SLE patients by spectratyping. Lupus 1997; 6:701-707.
38. Davey MP, Burgoine GA, Woody CN. TCRB clonotypes are present in CD4[+] T-cell populations prepared directly from rheumatoid synovium. Hum Immunol 1997; 55:11-21.
39. Vekony MA, Holder JE, Lee AJ et al. Selective amplification of T-cell receptor variable region species is demonstrable but not essential in early lesions of psoriasis vulgaris: Analysis by anchored polymerase chain reaction and hypervariable region size spectratyping. J Invest Dermatol 1997; 109:5-13.
40. Perreault C, Jutras J, Roy DC et al. Identification of an immunodominant mouse minor histocompatibility antigen (MiHA). T-cell response to a single dominant MiHA causes graft-versus-host disease. J Clin Invest 1996; 98:622-628.
41. Howell CD, Li J, Roper E et al. Biased liver T-cell receptor Vβ cell repertoire in a murine graft-versus-host disease model. J Immunol 1995; 155:2350-2358.
42. Danska JS, Livingstone AM, Paragas V et al. The presumptive CDR3 regions of both T cell receptor alpha and beta chains determine T-cell specificity for myoglobin peptides. J Exp Med 1990; 172:27-33
43. Caton AJ, Gerhard W. The diversity of the CD4[+] T-cell response in influenza. Immunol 1992; 4:85-90.
44. Michalides R, Wagenaar E, Groner B et al. Mammary tumor virus proviral DNA in normal murine tissue and non-virally induced mammary tumors. J Virol 1981; 39:367-376.
45. Traina VL, Taylor BA, Cohen JC. Genetic mapping of endogenous mouse mammary tumor viruses: Locus characterization, segregation, and chromosomal distribution. J Virol 1981; 40:735-744.
46. Kato T, Suzuki S, Sasakawa H et al. Comparison of the Jβ gene usage among different T-cell receptor Vβ families in spleens of C57BL/6 mice. Eur J Immunol 1994; 24:2410-2414.
47. Wettstein PJ. Murine minor histocompatibility antigens detected by helper T cells. Recognition of an endogenous peptide. J Immunol 1995; 154:2134-2143.
48. Goulmy E. Human minor histocompatibility antigens: New concepts for marrow transplantation and adoptive immunotherapy. Immunol Rev 1997;157:125-140.

CHAPTER 13

The Diversity and Characteristics of T-Cell Receptors Specific for Single Non-H2 Histocompatibility Antigens

Peter J. Wettstein and Sean L. Johnston

Many histocompatibility (H) antigens stimulate rejection of allografts and graft-versus-host disease (GVHD) following bone marrow transplantation in mammals. They can be subdivided into two groups:

1. alloantigens encoded by genes mapping to the major histocompatibility complex (MHC and H2 in mice) and
2. alloantigens encoded by numerous genes mapping to autosomes, sex-linked chromosomes, and mitochondria.

The alloantigens comprising the latter group are short immunogenic peptides that bind to the peptide binding sites of MHC class I or class II molecules and stimulate cytolytic T lymphocytes (CTL) or T helper (Th) cells, respectively, and have been referred to as either as non-H2 H antigens or minor H antigens.

The specificity of the T-cell response to minor H antigens is regulated by the binding of the peptides to MHC class I and class II molecules and recognition of these complexes by peptide-specific T-cell receptors (TCRs.) An understanding of this MHC/peptide/TCR interaction requires a molecular characterization of all three components. The identification of minor H peptides in humans and rodents is revealing the properties that enable them to bind to MHC molecules and distinguish donor peptides from those of recipients. However, the study of minor H antigen-specific TCRs has received only limited attention. In the case of minor H antigens with potential for stimulating graft versus leukemia responses, such information will be important for identifying peptide analogs with increased biological activity.

The antigen-specific component of the TCR is comprised of paired α and β chains. The α chain is encoded by Va, Ja, and Ca gene segments with rearrangement of the Va and Ja regions leading to 'N' nucleotide additions in the third complementarity determining region (CDR3). Similarly, the β chain is encoded by germline Vb, Db, Jb, and Cb genes with rearrangement between individual Vb genes and diversity (DB) segments followed by rearrangement to adjoin Jb genes. The Vb-Db-Jb rearrangements also involve the addition of 'N' nucleotides with variable usage of the $D\beta$ coding sequence leading to length and sequence variation in CDR3 sequences of the β chain. Antigen specificity of individual TCRs is the result of the choice of rearranged germline genes, the CDR3 sequences and the combination of the α and β chains. The recently published crystal structures of murine and human TCRs[1,2] have increased information on the α and β CDRs that contact class

Minor Histocompatibility Antigens: From the Laboratory to the Clinic, edited by Derry Roopenian and Elizabeth Simpson. ©2000 Landes Bioscience.

I:peptide complexes. CDR1α, CDR3α, and CDR3β potentially contribute to direct contact with class I-bound peptides.[2] The majority of studies of TCR recognition of MHC-bound peptides have involved immunogenic peptides derived from either viral antigens or model, foreign proteins, e.g., ovalbumin and xenogeneic cytochrome C. Because allelic differences of known minor H peptides are often limited to discrete amino acid substitutions,[3,4] it is likely that their self-allelic counterpart influences the TCR repertoire by acting as a ligand during negative and possibly positive thymic selection processes. Therefore, minor H antigen-specific TCRs may be more restricted compared to TCRs for more foreign peptides.

The investigation of TCR specificity has focused primarily on the study of T-cell clones selected in vitro. Direct studies of TCRs used in vivo have been more difficult because of contamination by non-specific bystander cells at sites of inflammation. The technique of spectratyping of in vivo samples has proved useful as TCR β chain transcripts can be amplified by RT-PCR to identify the *Vb* genes utilized and the length of CDR3β sequences.[5,6] Antigen-specific β chains are identified by the contraction of diverse CDR3β Vb-specific PCR products. However, this analysis is limited analysis of TCR β chains, in contrast to the sequence information obtained for both α and β chains using isolated T-cell clones.

Dichotomy Between Requirements for Allograft Rejection and CTL Generation

Differences between in vivo allograft rejection and CTLs derived following in vitro expansion can be explored by TCR analysis. One striking discrepancy between T-cell responses to minor H antigens involved in allograft rejection and the specific CTLs isolated after in vitro expansion is observed in donor-recipient combinations that involve multiple minor H antigens barriers. Skin allografts incompatible for multiple minor H antigens are rejected by naïve mice at speeds comparable to H2-incompatible grafts. However, only H2-disparate naïve mice generate antigen-specific CTLs in mixed lymphocyte cultures (MLC): minor H antigen-specific CTLs can only be generated after in vivo priming.[7] There are also differences in the specificities of the T cells involved: a multiple minor H-disparate skin allograft immunizes mice to reject in a second-set manner allografts disparate at any one of several single minor H antigens, while those same allografts prime for in vitro-detected CTLs specific for a much more restricted set of immunodominant minor H peptides.[8]

This dichotomy is further complicated by the differential requirements for CD4+ Th cells. The generation of minor H antigen-specific CTLs requires CD4+ Th cells,[9,10] but the requirement for CD4+ T-cells in the rejection of minor H incompatible skin allografts is not so clear. There are two disparate sets of data. In models involving immunodeficient recipients reconstituted with T cell subpopulations, minor H antigen-disparate skin grafts are greatly prolonged in the absence of CD4+ T cells.[11,12] However, we failed to detect a significant effect after antibody-mediated depletion of CD4+ T cells before grafting. It is possible to reconcile these observations: injections with relatively limited numbers of responder T cells may not fully reconstitute the immune response capability of the immunodeficient recipients, and the temporal delay between reconstitution of the immunodeficient recipients and grafting could attenuate the cytokine milieu that the responding T cells would encounter as they recognize graft antigens. During wound healing tumor necrosis factor-alpha (TNFα) is upregulated[13] and could stimulate Langerhans cell migration and cytokine and adhesion molecule expression by keratinocytes.[14] By separating transplantation and reconstitution, the absence this type of wound-healing cytokine could result in a dependence on CD4+ Th cells that is not required in recipients with fully functional CD8+ compartments.

H4-Specific TCRs Expressed by CTLs

To better understand the T cells responsible for graft rejection, we investigated the TCRs expressed by CTLs specific for the minor H antigen, H4, that stimulates CTLs as well as skin allograft rejection in vivo. The H4 antigen is defined by the response of C57BL/10 (B10) mice to its congenic strain B10.129-H4^b (21M).[15] The H47 antigen is detected in the reciprocal donor-recipient combination.[16] These antigens are not encoded by alternative alleles at the same locus but rather are encoded by distinct genes on chromosome 7 (Mendoza et al, submitted). CTL recognition of the H4 antigen is H2-K^b-restricted[17] and specific for a single immunogenic peptide that can be extracted from immunoprecipitated K^b molecules and separated as a single peak by reverse phase high performance liquid chromatography (HPLC).[18,19]

The immunodominance of H4 as a target antigen for CTLs is exemplified by the fact that it dominates over HY when female recipients are immunized with H4$^+$HY$^+$ 21M male spleen cells.[20] Gene mapping, strain distribution and biochemical analyses suggest that H4 is also the immunodominant "CTT-2" antigen preferentially recognized by C57BL/6 CTLs generated by immunizing and boosting with BALB.B spleen cells.[21,22,23]

A dual approach was taken to the analysis of TCRs expressed by H4-specific T cells:
1. sequencing of both alpha and beta chains expressed by H4-specific CTL clones selected in vitro and
2. spectratyping of β chains transcribed by CTLs that infiltrate H4-incompatible skin allografts at the time of rejection. The results of both of these analyses have been published.[24,26]

Studies of TCRs expressed by CTL clones utilized two sets of H4-specific clones derived from a number of individual B10 mice primed with 21M cells and B6 mice primed with CXBG spleen cells. The CXBG recombinant inbred strain was derived from an initial cross of B6 and BALB/C mice[27] and has been shown to express the CTT-2 antigen.[22,28] To insure TCR diversity, both sets of CTLs were cloned after only one stimulation in vitro with H4-positive stimulators. Alpha chain transcripts were reverse-transcribed and amplified by RT-PCR using Cα primers and either a consensus Vα primer[29] or primers that included restriction enzyme site sequences tailed with viral RNA polymerase promoter sequences as described previously.[30] β chain transcripts were amplified by RT-PCR with Cβ primers and either Vβ gene-specific primers or restriction enzyme site primers. No attempt was made to obtain full-length V gene sequences since somatic mutation is a rare event in TCR V genes.

The distribution of Va gene usage in these two panels of CTLs is indicated in Figure 13.1A. Although many Va genes were used by these clones, there was preferential usage of Va4 subfamily members with 53% of B10 anti-21M clones and 51% of H4-specific B6 anti-CXBG clones expressing one of these genes. Predominant usage of Va4 genes is not a typical characteristic of peptides presented by K^b molecules since CTLs specific for either the ovalbumin SIINFEKL peptide or the vesicular stomatitis virus (VSV) RGYVYQGL peptide do not exhibit such over-representation.[31,32] However, Va4 genes were exclusively expressed in a limited panel of CTL clones specific for a peptide derived from lymphocytic choriomeningitis virus and presented by D^b molecules.[33]

CTLs specific for H4 were also analyzed for CDR3α and CDR3β usage. Such analyses typically concentrate on length, amino acid composition, and presence of charged residues. CDR3α length was calculated as the number of amino acids between the terminal Vα Cys residue and the Jα GXG motif minus four amino acids, as proposed previously.[34] As overall length determines which parts of V and J segments will contact class I:peptide complexes, similarities in CDR3 length in clones specific for the same antigen are considered significant. Motifs of amino acid composition, such as the X-G-G-X-G-X-X motif found in beef insulin peptide/H2-A^d-reactive TCRs,[35] are also sought in a CDR3 region analysis. Finally,

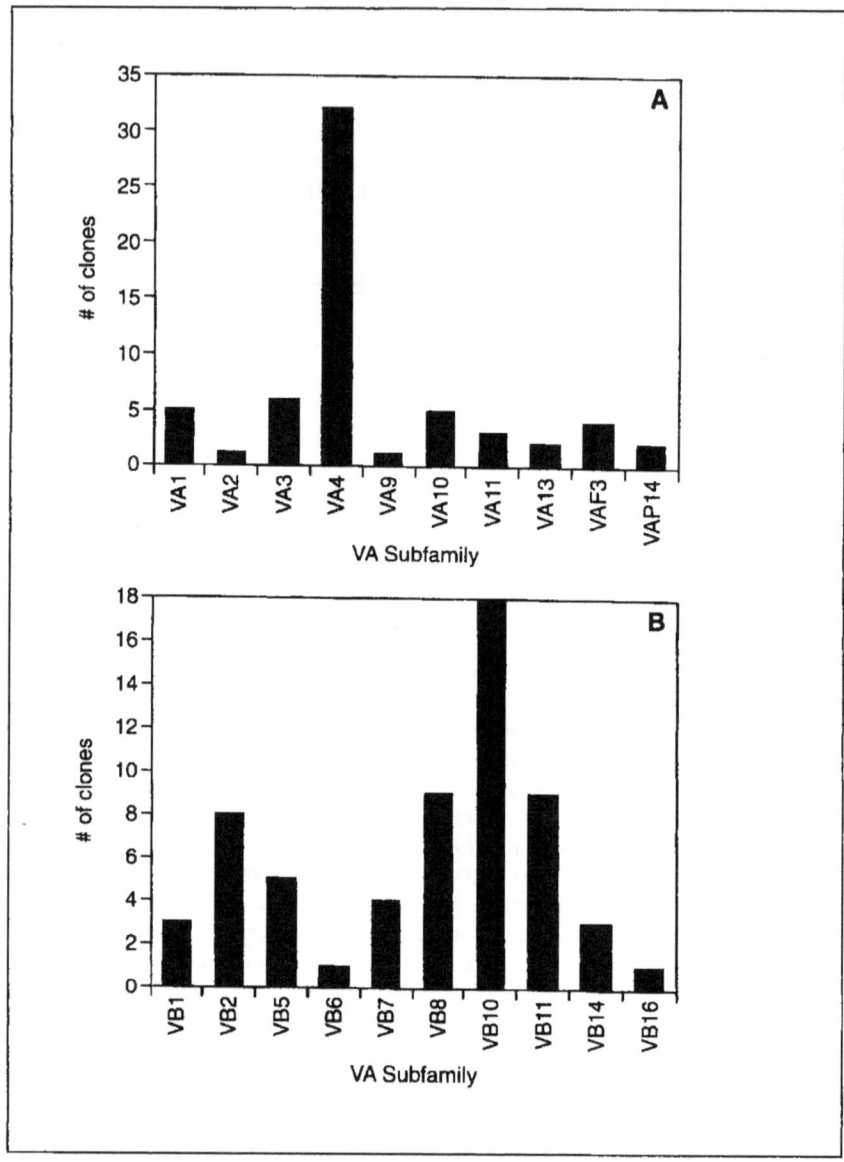

Fig. 13.1. Utilization of *Va* (A) and *Vb* (B) genes in cloned CTLs specific for the H4 peptide presented by K^b molecules. These clones were derived from B10 mice primed and boosted with H4-incompatible spleen cells[24] and B6 mice primed and boosted with CXBG spleen cells.[25]

as charged residues may form salt bridges with residues of class I:peptide complexes, the presence of these charged residues may also be significant. The CDR3 lengths in H4-specific α chains were distributed with a median and range comparable to that observed with randomly selected α chains. No apparent motifs were detected and the distribution of charged residues was consistent with that predicted from germline Jα sequences in that Lys residues are prominent in Jα sequences producing net positive charges.

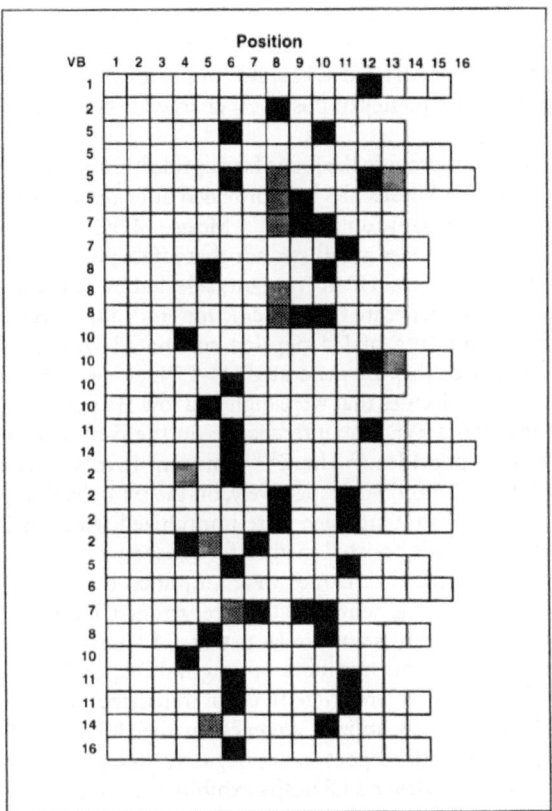

Fig. 13.2. Distribution of negatively and positively charged residues in the CDR3βs expressed by CTL clones specific for the H4 peptide in the context of K[b] molecules. These clones were derived from B10 mice primed and boosted with H4-incompatible spleen cells[24] and B6 mice primed and boosted with CXBG spleen cells.[25] Black and gray boxes represent negatively and positively charged amino acids, respectively. The CDR3βs depicted in this and the following figures start at the Cys residue in the terminal Vβ "CASS" element and end at the amino acid preceding the "GXG" motif in Jβ sequences.

Analysis of H4-specific TCRβ usage by cloned CTL lines showed oligoclonality with respect to *Vb* gene usage (Fig. 13.1B). The *Vb2, 5, 8, 10,* and *11* genes were significantly over-represented in many individuals, with more extensive use of *Vb8* genes in B6 anti-CXBG CTL clones. Although the represented *Vb* genes were limited in number, this panel did not exhibit the restriction in *Vb* usage of CTLs specific for the dominant VSV and ovalbumin peptides presented by K[b] molecules,[31,32] nor was there preferential combinations of particular *Va* and *Vb* genes. With respect to the CDR3β sequences, there were no apparent motifs or digression from a normal distribution of length. The distinguishing characteristic of these two panels of CDR3βs was the net negative charges of these sequences (Fig. 13.2). Negatively vs. positively charged amino acids were present at a 4:1 ratio, and the negatively charged amino acids were encoded by both significantly over-represented *Jb* gene segments that included Glu residues, and 'N' nucleotide additions. As can be seen from Figure 13.2, these charges tend to center on positions 6 and positions 10-11, the latter being contributed principally by utilized *Jb* genes. This net negative charge does not appear simply to be characteristic of K^b-restricted TCRs since the negative:positive ratios in panels of OVA and VSV-specific TCRs are ≈1:1 and ≈2:1, respectively. Based on the frequency of charged residues in these three panels of K^b-restricted CTLs, it is apparent that H4-specific CDR3βs have been selected for reduced frequency of positively charged residues in the N-Dβ-N segment of CDR3β.

MHC class I molecules provide the focus for peptide binding and recognition by TCRs, and crystal structures of class I molecules have provided information about the orientations of residues that impact this recognition process.[1,2] Mouse MHC class I mutants with altered residues on the α1 and α2 helices and β strands that affect peptide binding also facilitates this type of investigation. Natural H2-K^b mutants selected by skin allograft rejection differ in not only their abilities to generate CTL specific for H4,[19] but also in their ability to bind the H4 peptide.[18] The binding of H4 peptide correlated with the ability to generate H4-specific CTL. In view of this, we generated H4-CTL clones that were restricted by the K^{bm5} and K^{bm11} mutant molecules that were high and low H4 peptide binders, respectively. The K^{bm5} mutant involves a Tyr>Phe interchange on the β-pleated sheet at position 116 oriented toward bound peptides.[36] The K^{bm11} mutant involves Asp>Ser and Thr>Ala interchanges at positions 77 and 80, respectively, on the α1 helix that point toward bound peptides.[37] K^{bm5} and K^{bm11} mutant mice were immunized with H4-incompatible spleen cells and responder spleen cells boosted in vitro for subsequent CTL cloning. TCR α and β transcripts expressed by these CTL clones were amplified by RT-PCR and sequenced to identify utilized *V* and *J* gene segments and the intervening CDR3s.

Va gene usage in the two panels was similar to that of K^b-restricted CTLs but the relatively limited size of these panels and numbers of original responding mice preclude form conclusions. The K^{bm5}- and K^{bm11}-restricted β chain transcripts were similar to K^b-restricted β chains in terms of preferentially utilized *Vb* genes. Further, K^{bm5}- and K^b-restricted CDR3β regions were comparable with respect to net negative charges and positions of charged residues. However, K^{bm11}-restricted CDR3βs exhibited more focused negatively charged residues such that each unique sequence included one such amino acid at positions 10-11 with lesser frequency of occurrences at positions 4-6 (Fig. 13.3). Comparison of the distribution of negatively charged residues in K^{bm11}-restricted CDR3βs with those K^b- and K^{bm5}-restricted CTLs suggests that the amino acid substitutions associated with the K^{bm11} mutant may have altered this distribution.

The relatively strong net negative charges exhibited by H4-specific CDR3βs chains regardless of the *Vb* gene usage strongly suggest that such a charge is important for recognition of the K^b/H4 complex. The fact that CDR3βs of T-cell clones specific for other K^b restricted peptides do not display such strong net negative charges argues that these charges are not required simply for recognition of K^b molecules. These results suggested the H4 peptide included a positively charged residue in the carboxy terminus based on the reported crystal structures of human and mouse TCRs. This hypothesis was also based on the observation that K^{bm11}-restricted CDR3β sequences apparently required a negative charge in their carboxy ends. This charge may be required to compensate for the loss of the Asp residue at position 77 in the α1 helix. The interpretation is supported by the identification of H4 mimotopes included in a partially degenerate K^b-binding octapeptide library.[38] The library included peptides with Ser, Ile, Phe, and Leu fixed at p1, p3, p5, and p8, respectively,[39] and it was screened with a CTL clone that expresses Vβ8.1 and a CASSSGDTLYF CDR3β sequence. All H4 mimotopes included Gly at p2 and either Ile or Val at p4. The sixth position was more variable but obviously hydrophobic residues, such as Ile and Val, were preferred. Most importantly, all mimotopes included Arg or His at p7, consistent with our prediction based on CDR3β charge. Computer modeling of K^b:SGIVFVRL, based on the crystal structure of the K^b:RGYVYQGL complex, predicts that the Arg p7 residue orients directly out of the

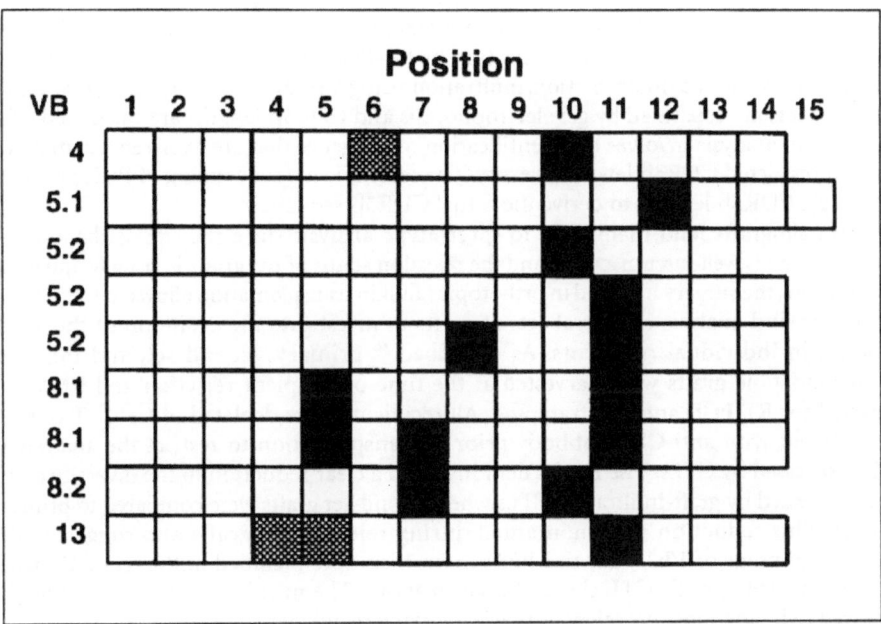

Fig. 13.3. Distribution of negatively and positively charged residues in the CDR3bs expressed by CTL clones specific for the H4 peptide in the context of mutant K^{bm11} molecules. Black and gray boxes represent negatively and positively charged amino acids, respectively.

complex and should be available to form salt bridges with opposing charges in CDR3β. Since the strength of these bonds is seven times that of van der Waal forces, these residues may be critical in stabilization of the complex of the TCR: K^b:H4 complex.

H4-Specific TCRs Expressed by Graft-Infiltrating CD8$^+$ T Cells

Although cloned T cells constitute important tools for identifying minor H peptides and for gaining TCR sequence information, they may be limited in their relevance to the in vivo T-cell response. For example, CTL clones may be derived from individuals undergoing either graft rejection or GVHD, but there is no assurance that these CTLs are actually involved in the in vivo T-cell response. Such CTLs characterized by having been expanded in vitro in the presence of specific APC. There is evidence in mice that skin allografts stimulate CTL specific for considerably more minor H antigens detected by subsequent allograft rejection than by CTLs expanded in vitro.[8]

Our studies of H4-specific TCRs were therefore extended to include the analysis of CD8$^+$ T cell that infiltrate H4-incompatible skin allografts in vivo. In vivo analyses are considerably more complicated than studies of TCRs expressed by cloned T cells if direct analysis is performed without T-cell expansion and cloning in vitro. However, the technique of spectratyping provides a method for identifying individual β chain sequences that are over-represented in sites of inflammation, i.e., graft sites.[5,6] This technique is based on observations that restriction in diversity of CDR3β lengths accompanies maturation of the T-cell response and precedes motif selection.[40] This contraction results in the loss of the Gaussian distribution of CDR3β length characteristic of both normal and polyclonally stimulated T cells expressing a particular *Vb* gene,[5] by the appearance of over-represented CDR3β lengths. This method has been applied to the study of T cells infiltrating tumors,[5,41]

T cells specific for recall antigens such as tetanus toxoid[6] and in graft-vs.-host studies (Chapter 12). Spectratyping involves the amplification by RT-PCR of βchain transcripts from RNA extracted from sites of inflammation/infiltration using Vβ- and Cβ-specific primer pairs. PCR products are separated by gel electrophoresis and CDR3β lengths are calculated. The first order of analysis involves the identification of *Vb* genes that are over-represented in a site with restricted CDR3β length diversity. The second is the sequencing of PCR products with single CDR3β lengths to derive the actual CDR3β sequences.

Skin allografts lend themselves to spectratype analysis since the site of the effector T cell response is well-circumscribed and the rejection status of the grafts is visually apparent. Furthermore, the surgery involved in orthotopic tail skin transplantation allows the sequential application and analysis of several sets of grafts to investigate the evolution of the T-cell response in individual recipients. As described,[26] primary, second-set, and third-set H4-incompatible grafts were harvested at the time of incipient rejection and RNA was extracted for RT-PCR and spectratyping. All recipients were depleted of CD4[+] T cells by pretreatment with anti-CD4 antibody prior to transplantation to restrict the analysis to TCRs expressed by CTLs. The results demonstrated a clear reduction in the diversity of *Vb* genes expressed by graft-infiltrating CTLs when second-set grafts were compared to primary grafts.[26] This reduction was maintained during rejection of grafts and suggested the preferential usage of *Vb2*, *Vb5*, and *Vb8* genes which were included in the set of *Vb* genes expressed by H4-specific CTL clones described above. The members of the *Vb8* subfamily appeared to be the most prevalent *Vb* genes in many recipients.

The reduction in *Vb* gene usage was accompanied by reduction in the diversity of CDR3β lengths such that individual PCR products could be sequenced to identify the predominant β transcripts. In fact, in a few cases single transcripts dominated the spectratypes from individual recipients. This sequencing could be accomplished in many cases by simply using Cβ primers; complexity of additional PCR products could be overcome by gel electrophoretic separation of multiple bands for sequencing or the use of *Jb*-specific primers for sequencing. The ease with which single-copy sequences were obtained contrasts with the reliance on *Jb*-specific primers for dissection of relatively complex PCR products in other spectratype analyses.[5,41,42] This difference may well be based on the fact that H4 is a single immunogenic peptide presented by K[b] molecules above, whereas T cells spectratyped in previous studies could be specific for multiple peptides presented by several class I and class II molecules.[5,41,42]

The CDR3β sequences identified did differ from their CTL clone counterparts in that they had a maximal length of 10 a.a. with a predominance of lengths of 9 and 10 a.a.: 25/47 β to their in vitro-derived counterparts in their overall, net negative charges and the relative positions of these residues within the CDR3β sequences (Fig. 13.4); the negative to positive charge ratio was ≈4:1 as observed with CTL clones in vitro. Although contraction in CDR3β length diversity predicts that such β chain transcripts are expressed by H4-specific CD8[+] cells, there is still a possibility that the sequenced transcripts are expressed by nonspecific bystander CD8[+] T cells at the site of rejection. However, a number of closely related sequences were found that among the CDR3β sequences expressed by graft-infiltrating CTLs in vivo and CTL clones in vitro.[26] First, a Vβ8.1-Jβ2.1 transcript with the CDR3β CASSEQGNYAEQFF sequence was identified in RNA derived from both second- and third-set H4 allografts on a single recipient. Second, this CDR3β sequence was found in a Vβ8.2-Jβ2.1 transcript from a different recipient. Third, a Vβ8.2-Jβ2.1 transcript with a Glu>Asp interchange at the fifth position was identified. Fourth, two additional Vβ8.2-Jβ.1 transcripts were identified in which the Glu residue at position five was altered but the Glu residue at position 11 was retained. Fifth, the CASSDQGNYAEQFF sequence expressed by graft-infiltrating DC8[+] T cells was identical at the amino acid level to CDR3β sequences

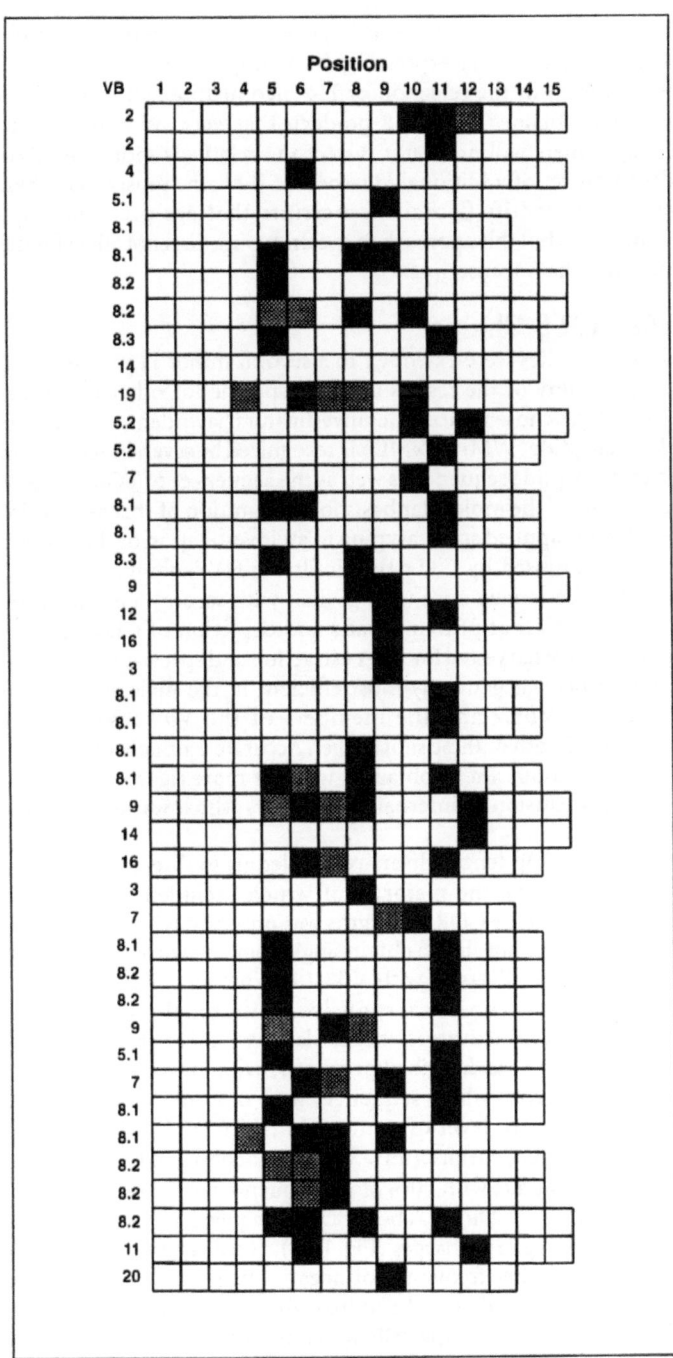

Fig. 13.4. Distribution of negatively and positively charged amino acids in CDR3βs expressed by CTLs that infiltrated second-set and third-set H4-incompatible skin allografts transplanted to B10 and B6 recipients.[26] Black and gray boxes represent negatively and positively charged amino acids, respectively.

expressed by three independent H4-specific generated CTL clones. These observations are consistent with the CD8$^+$ T cells expressing this family of closely related CDR3β sequences being directly involved in the rejection of H4-incompatible allografts. However, CDR3β sequences were only obtained from PCR products that were easily analyzed; further analyses with *Jb*-specific primers will no doubt dissect the relatively more complex products and allow the identification of additional H4-specific β chain transcripts. These experiments have allowed us to identify β chain transcripts that are preferentially expressed by graft-infiltrating CTLs, but this approach to spectratyping does not allow for the identification of the TCR partner chain transcripts.

HY-Specific TCR β Chains

These in vivo studies were extended to a second minor H peptide system to evaluate the general applicability of the results to TCRs specific for other minor H peptides. The male antigen (HY) was chosen since other investigators have identified an immunodominant D^b restricted nonapeptide (WMHHNMDLI) recognized by several T-cell clones.[43] Knowledge of both the natural peptide sequence as well as the sequences of TCR α and β chains should facilitate the analysis of the molecular basis for recognition of this peptide in the context of H2-D^b molecules. We applied spectratyping to an investigation of the diversity and identity of TCR β chains expressed by CTLs that infiltrate HY-incompatible skin grafts on H2^b female recipients.[44] B6 female mice were grafted with second- and third-set HY-disparate grafts after their rejection of primary grafts and depletion of CD4$^+$ T cells. Grafts in the process of rejection were harvested for RNA extraction and spectratype analysis. Amplification of β chain transcripts was generally most efficient in the majority of recipients with V*b* primers specific for V*b*5.2 and the members of the V*b8* subfamily, suggesting the over-representation of both of these subfamilies. Accurate estimates of the levels of expression of different V*b* genes could not be obtained without more rigorous control of the RT-PCR conditions and the inclusion of internal competitors which was beyond the scope of these experiments.

Restriction in CDR3β length diversity enabled us to directly sequence a considerable number of PCR products, the majority of which included V*b8* subfamily members. Although reduction to single CDR3β lengths was observed with Vβ5.2 chain transcripts in many individuals, we were not able to obtain single copy sequence from any of these products. The restriction in length diversity with Vβ5.2 chain transcripts indicated that the CTLs expressing these sequences were responsive to nominal antigen rather than either superantigens or polyclonal activators since the latter would be expected to result in Gaussian CDR3 length distributions as described previously with human Th cells.[6]

However, the ease with which sequences were obtained from Vβ8 chain transcripts demonstrated the relative level of contraction of CDR3β length diversity based on over-representation of single transcripts. Sequence analysis of CDR3βs revealed that their lengths varied from 7-11 a.a. with 50% of the sequences having lengths of 9 a.a., similar to the results with H4-specific CDR3βs. These CDR3β sequences were also similar to H4-specific sequences in their net negative charges (Fig. 13.5). However, in contrast to H4, HY-specific sequences included a higher frequency of negatively charged residues at positions 5-6 in addition to the residues encoded by the utilized Jb gene segments. Further, there were fewer positively charged amino acids especially in sequences derived from third-set grafts where no such residues were found; here the ratio of negative:positive charges was ≈6:1. A major determinant in the presence of Asp residues at position 5 was the apparent retention of the 'GAT' codon that immediately follows the nucleotide sequence encoding the 'CASS' element in germline V*b8* genes.[45] Therefore, inclusion of negatively charged amino acids in these positions was based on both 'N' nucleotide additions as well as differential rearrangements

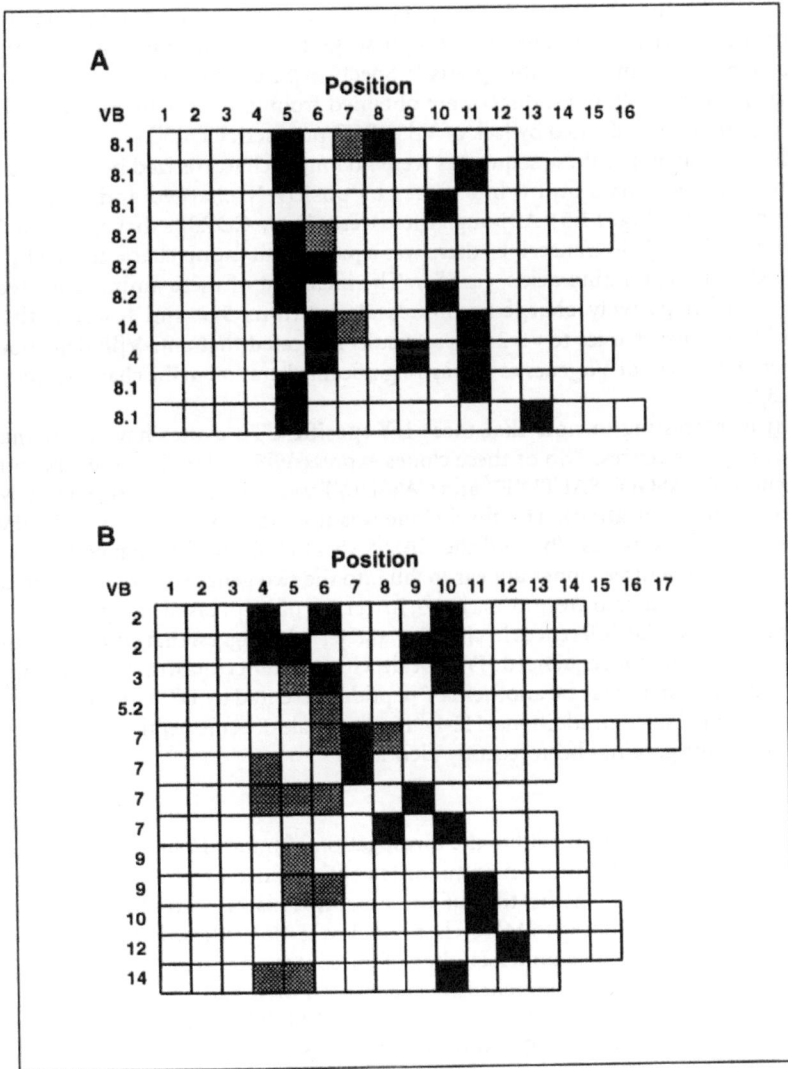

Fig. 13.5. Distribution of negatively and positively charged amino acids in CDR3βs expressed by CTLs that infiltrated HY-incompatible skin allografts transplanted to B6 female recipients.[44] (A) The presented CDR3β sequences were derived from second- and third-set HY-disparate allografts placed on normal recipients. (B) The depicted CDR3β sequences were derived from second-set HY-disparate allografts placed on recipients that had been pre-treated with anti-Vβ8 monoclonal antibody and confirmed to be depleted of Vβ8⁺ T cells. The charged residues are represented as in Figs 13.2-13.4.

involving the *Vb8* genes. As in the case of the H4 response, it was possible to identify a CDR3β sequence present in both second- and third-set grafts on an individual mouse, strongly implicating the CD8⁺ T cells expressing this beta chain in the HY-specific rejection process.

The predominance of *Vb8* gene expression suggested that in vivo depletion of Vβ8⁺ T cells with anti-Vβ8 antibody could slow HY-incompatible graft rejection as observed with skin and cardiac allografts that express the H2-L^d alloantigen that preferentially stimulates Vβ8$^+$ CTL in vitro.[46] However, treatment of recipients with anti-Vβ8 antibody after primary graft rejection and prior to second-set grafting had no effect on the survival times of second-set HY-incompatible grafts.[44] Spectratype analysis of these second-set grafts revealed that non-Vβ8$^+$ products were obtained from grafts on all recipients depleted of Vβ8$^+$ T cells and confirmed by lack of Vβ8$^+$ PCR products. CDR3β sequencing of non-Vβ8 products revealed that these sequences were no longer characterized by negatively charged residues at positions 5 and 6 but rather by positively charged and polar residues at these positions (Fig. 13.5). Although there were clearly CDR3βs that exhibited the charge characteristics of Vβ8 transcripts, there was a prominent group characterized by positively charged and polar amino acids, e.g., Ser. The inclusion of these amino acids reduced the number of negatively charged residues at positions 5-7 and lowered the ratio of negative:positive charges to ≈3:2. These results indicated that anti-Vβ8 depletion resulted in alternative usage of *Vb* genes as well as a significant alteration in the charge characteristics of CDR3βs.

It is interesting to note that three HY-specific CTL clones have been analyzed for β transcript sequences. Two of these clones express *Vb8* subfamily members with CDR3β sequences of CASGDNSAETLYF[47] and CASSDLVEVFF (O. Lantz, J. Ridge, and P. Matzinger, personal communication). The third clone was used to identify the WMHHNMDLI HY peptide, and it expresses Vβ7 and the CASSSGNTLYF CDR3β sequence.[44] Therefore, the β chains of these three clones appear to fall into the two general classes expressed by CTLs infiltrating HY-disparate grafts, both in terms of expressed *Vb* genes and CDR3β characteristics. The differences in charge characteristics suggest that either these two groups of TCRs recognize the sequenced HY peptide with different conformations, alignments and/ or contacts, or there may be another HY peptide presented by D^b. The latter possibility has been confirmed since an additional H2-D^b binding male-specific peptide encoded by the *Smcy* Y chromosome gene has been recently identified.[48]

Conclusion

The limited attention paid to the investigation of minor H specific TCRs is striking in comparison to the identification of minor H peptides currently burgeoning in human and mouse studies. The identification of these peptides is essential for understanding the molecular basis for binding to MHC class I or class II molecules and recognition by effector T cells. However, a detailed understanding of the control of specificity of the T-cell response to these peptides requires comprehensive analysis of TCR α and β subunits and, in particular, the regions that contact and bind to non-MHC:MHC complexes. Results of detailed investigations of TCRs specific for minor H peptides will be important at both basic and clinical levels. There is accumulating evidence that donor minor H peptides differ from their homologous or allelic counterparts by a limited number of amino acid substitutions.[3,4] These alternative peptides differ in their abilities to be recognized by specific CTLs; both can be recognized but appear to differ in affinity of their interactions with class I molecules and/or TCRs. These peptides are likely to be involved in thymic selection.

Clinically, minor H peptides stimulate the rejection of MHC-matched allografts, so that general immunosuppression is required to thwart this rejection process Our studies of the TCRs specific for minor H peptides were aimed at estimating the diversity of *Vb* gene usage, to assess the feasibility of anti-Vβ antibody-mediated depletion for immuno-suppression modality. The results of our initial studies, both in vitro and in vivo, suggest that minor H specific TCRs utilize diverse sets of *Vb* genes and such therapy would be unsuccessful. An alternative view is derived from a study of minor H specific TCRs expressed by a panel of HA-1-specific CTLs from several individuals suffering chronic

GVHD.[49] These CTLs all utilized a single *Vb* gene with diverse usage of *Va* genes. The basis for the difference between this and our observation may involve differences in

1. the minor H peptides themselves,
2. the extent of in vitro selection, or
3. skin allograft rejection versus chronic GVHD.

Although our experiments have involved primary, second-set, and third-set rejections, the experimental time periods are less than two months, short in comparison to the time periods of clinical chronic GVHD.

Clinical utility of minor H antigens may be found in graft versus leukemia (GVL) responses. Some minor H antigens appear to be restricted to cells of the hematopoietic lineage and could serve as effective targets of GVL responses with reduced risk of GVHD. Knowledge of the identity of the target peptides as well as TCRs specific for them should facilitate improvement in treatments. The identification and sequencing of TCR α and β chains would allow their transfection into immortalized T cells for possible identification of minor H peptides. Further, these cloned cDNAs may also be used to transfect stem cells and mature T cells used in bone marrow transplantation to increase the frequency of specific CTLs that have GVL function. In summary, detailed investigations of minor H peptides and the TCRs specific for them are required to understand the molecular bases for the specificity of non-MHC-specific T-cell responses and to develop methods for either suppressing or augmenting these responses in the clinical setting.

References

1. Garcia KC, Degano M, Stanfield RL et al. An αβ T-cell receptor structure at 2.5A and its orientation in the TCR-MHC complex. Science 1996; 274:209-219.
2. Garboczi DN, Ghosh P, Utz U et al. Structure of the complex between human T-cell receptor, viral peptide and HLA-A2. Nature 1996; 384:134-141.
3. Wang W, Meadows LR, den Haan JMM et al. Human H-Y: A male-specific histocompatibility antigen derived from the SMCY protein. Science 1995; 269:1588-1590.
4. Mendoza LM, Paz P, Zuberi A et al. Minors held by majors: The H13 minor histocompatibility locus defined as a peptide/MHC class I complex. Immunity 1997; 7:461-472.
5. Puisieux I, Even J, Pannetier C et al. Oligoclonality of tumor-infiltrating lymphocytes from human melanomas. J Immunol 1994; 153:2807-2818.
6. Currier JR, Deulofeut H, Barron KS et al. Mitogens, superantigens and nominal antigens elicit distinctive patterns of TCRB CDR3 diversity. Human Immunology 1996; 48:39-51.
7. Bevan MJ. H2 restriction of cytolysis after immunization of minor H congenic pairs of mice. Immunogenetics 1976; 3:177-184.
8. Nevala WK, Paul C, Wettstein PJ. Reduced diversity of CTLs specific for multiple minor histocompatibility antigens relative to allograft rejection in vivo. J Immunol 1997; 158:1102-1107.
9. Roopenian DC, Anderson PS. Generation of helper cell-independent cytotoxic T lymphocytes is dependent upon L3T4+ helper T cells. J Immunol 1988; 141:391-397.
10. Wettstein PJ and Korngold R. T-cell subsets required for the generation of CTL activity directed toward single and multiple histocompatibility antigens. Transplantation 1991; 54:296-307.
11. Roopenian DC, Anderson PS. Adoptive immunity in immune-deficient scid/scid mice. Transplantation 1988; 46:899-904.
12. Rosenberg AS, Mizuochi T, Sharrow SO et al. Phenotype, specificity, and function of T-cell subsets and T cell interactions involved in skin allograft rejection. J Exp Med 1987; 165:1296-1315.
13. Hubner G, Brauchle M, Smola H et al. Differential regulation of pro-inflammatory cytokines during wound healing in normal and glucocorticoid-treated mice. Cytokine 1996; 8:548-556.

14. Kondo S, Sauder DN. Tumor necrosis factor (TNF) receptor type 1 (p55) is a main mediator for TNF-α-induced skin inflammation. Eur J Immunol 1997; 27:1713-1718.
15. Snell GD, Stevens LC. Histocompatibility genes of mice. III. H-1 and H-4, two histocompatibility loci in the first linkage group. Immunology 1961; 4:366-379.
16. Davis AP, Roopenian DC. Complexity at the mouse minor histocompatibility locus H-4. Immunogenetics 1990; 31:7-12.
17. Wettstein PJ, Frelinger JA. H2 effects on cell-cell interactions in the response to single non-H2 alloantigens. III. Evidence for a second *Ir* gene system mapping to the H2K and H2D regions. Immunogenetics 1980; 10:211-226.
18. Wettstein PJ, van Bleek GM and Nathenson SG. Differential binding of a minor histocompatibility antigen peptide to H-2 class I molecules correlates with immune responsiveness. J Immunol 1993; 150:2753-2760.
19. Wettstein PJ, Melvold RW. H2 effects on cell-cell interactions in the response in the response to single non-H2 alloantigens. VI. H2-K^b mutants differentially regulate the immune response to the H4.2 alloantigen. Immunogenetics 1983; 17:109-123.
20. Wettstein P. Immunodominance in the T-cell response to multiple non-H2 histocompatibility antigens. II. Observation of a hierarchy among dominant antigens. Immunogenetics 1986; 24:24-31.
21. Nevala WK, Wettstein PJ. H4 and CTT-2 minor histocompatibility antigens: Concordant genetic linkage and migration in two-dimensional peptide separation. Immunogenetics 1996; 44:400-404.
22. Nevala WK, Wettstein PJ. The preferential CTL response to immunodominant minor histocompatibility antigen peptides. Transplantation 1996; 62:283-291.
23. Strausbauch MA, Landers JP, and Wettstein PJ. Mechanism of peptide separations by solid phase extraction capillary electrophoresis at low pH. Anal Chem 1996; 68:306-314.
24. Johnston SL, Graham D, Wettstein PJ. Diversity of alpha and beta subunits of T-cell receptors specific for the H4 minor histocompatiblity antigen. Immunogenetics 1997; 46:17-28.
25. Johnston SL, Wettstein PJ. T-cell receptor diversity in CTLs specific for the CTT-1 and CTT-2 minor histocompatibility antigens. J Immunol 1997; 159.
26. Johnston SL, Borson ND, Wettstein PJ. Spectratyping of TCRs expressed by CTL-infiltrating minor histocompatibility antigen-disparate allografts. J Immunol 1997; 159:5233-5245.
27. Bailey DW. Recombinant inbred strains. Transplantation 1971; 11:325-327.
28. Wettstein PJ, Bailey DW. Immunodominance in the immune response to "multiple" histocompatibility antigens. Immunogenetics 1982; 16:47-58.
29. Moonka D and Loh EY. A consensus primer to amplify both α and β chains of the human T cell receptor. J Immunol Meth 1994; 169:41-51.
30. Johnston SL, Strausbauch M, Sarkar G et al. A novel method for sequencing members of multi-gene families. Nucl Acids Res 1995; 23:3074-3075.
31. Kelly JM, Sterry SJ, Cose S et al. Identification of conserved T cell receptor CDR3 residues contacting known exposed peptide side chains from a major histocompatibility complex class I-bound determinant. Eur J Immunol 1993; 23:3318-3326.
32. Imarai M, Goyarts EC, van Bleek GM et al. Diversity of T cell receptors specific for the VSV antigenic peptide (N52-59) bound by the H2-K^b class I molecule. Cell Immunol 1995; 160:33-42.
33. Aebischer T, Oehen S, Hengartner H. Preferential usage of VA4 and VB10 T-cell receptor genes by lymphocytic choriomeningitis virus glycoprotein-specific H2D^b-restricted cytotoxic T cells. Eur J Immunol 1990; 20:523-531.
34. Rock EP, Sibbald PR, Davis MM et al. CDR3 length in antigen-specific immune receptors. J Exp Med 1994; 179:323-328.
35. Wither J, Pawling J, Phillips L et al. Amino acid residues in the T-cell receptor CDR3 determine the antigenic reactivity patterns of insulin-reactive hybridomas. J Immunol 1991; 146:3513-3522.
36. Pease LR, Horton RM, Pullen JK et al. Structure and diversity of class I antigen presenting molecules in the mouse. Critical Reviews in Immunology 1991; 11:1-32.

37. Geliebter J, Nathenson SG. Microrecombinations generate sequence diversity in the murine major histocompatibility complex: analysis of the K^{bm3}, K^{bm4}, K^{bm10}, and K^{bm11} mutants. Mol and Cell Biol 1988; 8:4342-4352.

38. Strausbauch MA, Nevala WK, Roopenian DC et al. Identification of mimotopes for the H4 minor histocompatibility antigen. Intl Immunol 1998; 10:421-434.

39. Gavin MA, Dere B, Grandea AG, III et al. Major histocompatibility complex class I allele-specific peptide libraries: Identification of peptides that mimic an HY T-cell epitope. Eur J Immunol 1994; 24:2124-2133.

40. McHeyzer-Williams MG, Davis MM. Antigen-specific development of primary and memory T cells of vivo. Science 1995; 268:106-111.

41. Caignard A, Dietrich P, Morand V et al. Evidence for T-cell clonal expansion in a patient with squamous cell carcinoma of the head and neck. Cancer Research 1994; 54:1292-1297.

42. Dietrich PY, Caignard A, Lim A et al. In Vivo T-cell clonal amplification at time of acute graft-versus-host disease. Blood 1994; 84:2815-2820.

43. Greenfield A, Scott D, Pennisi D et al. An HY-D^b epitope is encoded by a novel mouse *Y* chromosome gene. Nature Genetics 1996; 14:474-478.

44. Johnston SL, Wettstein PJ. Spectratyping of TCR expressed by CTL infiltrating male antigen (HY)-disparate allografts. J Immunol 1998; 159:5233-45.

45. Chou HS, Anderson SJ, Louie MC et al. Tandem linkage and unusual RNA splicing of the T-cell receptor beta-chain variable-region genes. Proc Natl Acad Sci USA 1987; 84:1992-1996.

46. Cosman D, Fanger N, Borges L et al. A novel immunoglobulin superfamily receptor for cellular and viral MHC class I molecules. Immunity 1997; 7:273-282.

47. Uematsu Y, Ryser S, Dembic Z et al. Transgenic mice the introduced functional T cell receptor β gene prevents expression of endogenous β genes. Cell 1988; 52:831-841.

48. Markiewicz MA, Girao C, Opferman JT et al. Long-term T cell memory requires the surface expression of self-peptide/major histocompatibility complex molelcules. Proc Natl Acad Sci USA 1998; 95:3065-3070.

49. Goulmy E, Pool Jvan den Elsen PJ. Interindividual conservation of T-cell receptor β chain variable regions by minor histocompatibility antigen-specific HLA-A*0201-restricted cytotoxic T-cells clones. Blood 1995; 85:2478-2481.

50. Goulmy E. Human minor histocompatibility antigens. Curr Opin Immunol 1996; 8:75-81.

CHAPTER 14

Infectious Tolerance with CD4 Monoclonal Antibodies: A Role for Minors in Linked Suppression

Matt Wise, Frederike Bemelman, Stephen Cobbold and Herman Waldmann

O ne of the major goals of therapeutic immunosuppression is to use short-term therapy to achieve long-term tolerance. Short courses of treatment with antibodies specific for CD4 molecules on T cells can create peripheral tolerance in a mature rodent immune system to skin, heart or marrow grafts. Tolerance can be elicited to minor histocompatibility antigens (minors) even if they have been indirectly processed by the graft recipient. As discussed below, the tolerance so obtained is dominant, "infectious" and transferable from one animal to another. Tolerance to one set of minors can extend to other minors or even major histocompatibility complex (MHC)-encoded alloantigens (majors) if both types of antigens are expressed on the same grafted cells. This property of "linked suppression" may have considerable therapeutic potential in both allogeneic and xenogeneic transplantation in which one tolerises to sets of minors present on the donor tissue before the therapeutic graft is attempted. Immunodominant minor antigens may be particularly valuable in their exploitation as "vaccines" to achieve linked suppression.

Introduction

Transplantation of bone marrow and other organs is constrained by fierce rejection processes, which currently require control with immunosuppressive drugs, each with unwanted side effects. The ideal immunosuppressant would be one that could be given short-term, with few side effects and yet yield long-term benefits; in other words one that can achieve immunological tolerance. As T cells are crucial for transplant rejection, the goal should be to achieve long-term T-cell tolerance. Ideally, this should be achieved without wholesale ablation of the peripheral T-cell population given the fact that such T cells are not efficiently replenished in adult life by thymocyte progenitors.

In 1985 we observed that skin transplants given under cover of lytic anti-CD4 and CD8 monclonal antibodies (Mabs) survived long-term even though peripheral T cells had returned to normal levels.[1] In 1986 both we and David Wofsy demonstrated that short-term treatment with anti-CD4 Mabs could produce long-term tolerance to co-administered antigens.[2-4] Subsequently it was shown that non-lytic CD4 antibodies were also able to induce tolerance.[5-10] Not only could CD4 Mabs impose this on a naive immune system, but also on animals that had previously been immunized, or on animals actively rejecting their grafts.

Minor Histocompatibility Antigens: From the Laboratory to the Clinic, edited by Derry Roopenian and Elizabeth Simpson. ©2000 Landes Bioscience.

Evidence for Dominant Tolerance Mediated by CD4 T Cells

In all transplants that we have studied to date, including skin, heart and bone marrow, animals tolerant as a result of non-lytic CD4 mAb therapy have CD4 T cells with a capacity for dominant regulation of the rejection process (Table 14.1). It is difficult to break tolerance in such animals by infusion of naive T cells unless CD4 cells from the tolerant host are first depleted (a phenomenon we have termed resistance). Animals tolerant to one set of minors or majors (A-type) are impaired in their capacity to reject grafts from (A x B) F1 donors in which B contributes "third party" minors or majors. This phenomenon is referred to as linked suppression. CD4 T-cells from tolerant animals can also transfer tolerance to naïve recipients (transferable suppression). For example, adoptive transfer of T cells from mice tolerant to an A-type vascularised heart graft will impose tolerance even on a fully immuno-competent mouse to the extent that the hosts accept an MHC-mismatched A-type heart graft not only from the same donor, but also from an (A x B) F1 hybrid (Table 14.1). Transfer of T cells from these tolerant secondary hosts to tertiary recipients enabled acceptance of the third-party B-type heart grafts. As no prophylactic immunosuppression was given to these secondary or tertiary recipients we concluded that dominant regulatory processes were responsible for the spread of tolerance, and that all were in-built within the immune system and revealed only as a result anti-CD4 mAb therapy of the primary recipient.

How is Dominant Tolerance Maintained Life-Long?

Once dominant tolerance has been imposed on an immune system how does it remain life-long? Does the first cohort of regulatory CD4 T cells impose tolerance throughout life, or is this function passed on to newly recruited regulatory T cells? Clear evidence in support of the latter possibility has been obtained from experiments demonstrating that donor cells confer host T cells with the capacity to tolerize. (Table 14.2). In one case this infectious tolerance was associated with linked-suppression showing that these events are coupled and part of the same process.[17] The demonstration that tolerance is lost once T cells are "parked" in mice that lack the appropriate antigen[12,18] suggests that antigen must be available to sustain the regulators, and to recruit new regulators by the infectious tolerance process. This process and the finding that the depletion of CD4[+] T-cells in primed, tolerant hosts results in a breakdown of tolerance[18] suggest that some T cells competent to reject grafts remain within tolerant animals, but are subdued by CD4[+] T-cell regulators that are continuously interacting with antigen.

Several groups have now published on a range of antibodies that can induce tolerance without the need to deplete T cells. However, very few have searched for evidence of dominant tolerance. The central mechanism seems to be that the different antibodies interfere with secondary signals to T cells and that T cells become tolerant by default signalling through the TCR (signal 1). We prefer the concept that tolerogenic antibodies create a cease-fire in which the immune system cannot aggress the graft but can still be induced by graft antigens to elicit regulatory cells.[19] Once immunosuppression is removed the balance of regulators vs. effectors determines whether rejection or tolerance will ensue. As a corollary, it is possible that conventional immunosuppressants have not been successful agents for tolerance induction because they suppress both regulator and effector cells.

How is Dominant Tolerance Achieved?

The mechanism by which CD4 T cells suppress allograft rejection is unresolved, although IL4 has been implicated in one model,[16] but not in another.[18] Despite observing that suppression could be partly inhibited by neutralising anti-IL4 Mabs, no clear evidence could be found for immune deviation towards a Th2 phenotype by limiting dilution analysis.[20] Plain et al[21] also observed that Th2 cytokines were downregulated in animals tolerised to

Table 14.1. Evidence in support of dominant tolerance mediated by CD4 T cells

Observation	Reference
Lymphocyte infusions demonstrate resistance to the breakdown of tolerance	11,12
Resistance is mediated by CD4 T cells from tolerant hosts	11,12
Linked suppression	
Tolerance to graft A results in acceptance and tolerance to (A x B)F1 graft	13,14
CD4 T cells transferred from mice tolerized to MHC-mismatched A-type heart grafts suppress immunocompetent recipients from rejecting hearts from A donors and from heterozygous (A x B)F1 donors.	14
Naive hosts that receive T cells from the above tolerant animals accept B-type hearts	14
Mice tolerized to low doses of "minor"-mismatched A donor marrow grafts accept skin from (A x B)F1 donors.	17
Transferable suppression	
CD4 T cells, but not CD8 T cells, from tolerant animals suppress naive T cells	15,16,17
CD4 T cells from primed, then tolerized, animals suppress both naive and primed T cells	18
CD4 T cells from low-marrow dose tolerized mice transfer linked suppression to naive mice	17

allografts with a non-depleting CD4 Mab. It is still possible that dominant tolerance is a passive process involving inhibition by anergic CD4 T cells (the "Civil Service Model"[22]). Linked-suppression suggests that regulatory CD4 T cells and the potentially aggressive T cells they suppress, must come into close contact within the same microenvironment, if not onto the same antigen presenting cell (APC). Although dominant tolerance is characterised by regulatory CD4 T cells, the role that the APC itself plays remains to be elucidated, and may yet turn out to be equally important.

Therapeutic Application

The studies described above suggest that life-long dominant peripheral T-cell tolerance can be imposed not only on a naive immune system but even on one that has been previously exposed to antigens. Dominant tolerance to one set of antigens can spread to other sets if they are on the same tissue (linked suppression). How might such a powerful mechanism be exploited therapeutically?

T-cell recognition of transplanted tissue antigens is focussed on APCs, which may originate either from the graft itself (direct recognition) or from the host (indirect recognition). It had been considered for some time that the direct route of antigen presentation was the primary pathway through which graft rejection occurred. However donor skin from MHC Class I and II knockout mice is rejected as rapidly as skin transplants

Table 14.2. Evidence in support of infectious tolerance

Observation	Reference
Naive T cells become tolerant on transfer to a tolerant host	15
Tolerant T cells endow tolerance to heart grafts onto a secondary naive host	14
Tolerance can be transferred through 10 consecutive passages into naive hosts by infusion of spleen cells	14
Tolerant T cells from the spleens of antigen-boosted marrow-tolerant donors impose tolerance on secondary naive hosts	17

from wild type mice,[23] illustrating that at least for skin, the indirect pathway is as powerful as the direct route in initiating rejection. This is not true of vascularised organ grafts where presentation of reprocessed antigen on host APCs alone can lead to indefinite graft survival in some strain combinations.[24,25] We had also made the observation that tolerance induced by non-depleting anti-CD4 mAbs extended to minors presented through the indirect pathway.[13] Thus when (CBA/Ca x BALB/c) F1 mice ($H2^{k \times d}$) were transplanted with minor-mismatched B10.BR skin ($H2^k$) under cover of CD4 antibody, these recipients accepted not only a second B10.BR but also a B10.D2 ($H2^d$) skin transplant. We reasoned that if tolerance to minors could operate through the indirect route, then linked suppression may show the same features.

Using the same (CBA/Ca x BALB/c) F1 mice ($H2^{kxd}$) as recipients, we induced tolerance with non-depleting antibody to either B10.BR ($H2^k$) or B10.D2 ($H2^d$) in thymectomised (to demonstrate suppression is a peripheral event) or euthymic animals (Table 14.3). Recipients tolerant of B10.BR showed linked suppression to (B10.BR x CBK) F1 ($H2^k$) or (B10.BR x AKR) F1 ($H2^k XH2K^b$) skin grafts, where the third party antigen is an MHC class I disparity (suppression is indicated by delay in rejection) or multiple minors (these grafts survived indefinitely) respectively (Table 14.3). Next we transplanted the same F1 donor grafts onto the recipients tolerant of B10.D2. As these animals also showed evidence of linked suppression, we must conclude the regulatory CD4 T cells generated through recognition of B10.D2 minors ($H2^d$) represented on host $H2^k$ were preventing rejection of the $H2^k$ F1 skin grafts. Indeed the median survival time (MST) of the F1 donor was similar irrespective of whether the tolerizing graft was B10.BR or B10.D2.

The observation that linked suppression can operate through the indirect pathway provides a exciting new opportunity to exploit dominant mechanisms of tolerance, which can be applied to both allogeneic and xenotransplantation. Currently the variability in donor organ supply means that both the timing of transplantation and the incompatibilities between donor and host are unknown. One is then required to give the immunosuppression or tolerizing antibody treatment at the time of transplantation in the hope that regulatory T cells will predominate over lymphocytes that can initiate rejection. Clearly the chances of this strategy succeeding will be reduced when the transplant is very immunogenic. Knowledge of the molecular identity of common dominant minors would allow one to use some of these as 'vaccines' to promote tolerance through development of regulatory T cells. The observation that linked suppression is effective when antigen is presented by the indirect

Table 14.3. Linked suppression of allogeneic skin grafts

Tolerizing graft	F1 Graft	MST (days)	n	p
Recipients thymectomized (CBA/Ca x BALB/c) F1				
B10.BR	(CBA/Ca x CBK)	12	8	
				<0.005
B10.BR	(B10.BR x CBK)	17	9	
B10.D2	(CBA/Ca x CBK)	14	6	
				<0.008
B10.D2	(B10.BR x CBK)	20	15	
Recipients euthymic (CBA/Ca x BALB/c) F1				
B10.BR	(CBA/Ca x AKR)	21	8	
				<0.00008
B10.BR	(B10.BR x AKR)	>56	9	
B10.D2	(CBA/Ca x AKR)	25	8	
				<0.00002
B10.D2	(B10.BR x AKR)	>56	12	

(CBA/Ca x BALB/c) F1 mice ($H2^{kxd}$) were thymectomised at five weeks of age or were genetically euthymic. At six to eight weeks of age mice were grafted with either B10.BR ($H2^k$) or B10.D2 ($H2^d$) skin and given three 2 mg doses of YTS 177.9.6 and YTS 105 (non-depleting anti-CD4 and anti-CD8 mAbs) over a one week period. After 60 days the animals were regrafted with: (CBA/Ca x CBK) F1 or (CBA/Ca x AKR) skin, which was rejected; or (B10.BR x CBK) F1 or (B10.BR x AKR) F1 grafts, which showed prolonged survival.

pathway means that minors may be given in any form predisposed to undergo reprocessing on host MHC class II, for example as recombinant proteins.

Linked suppression may well explain the well-documented random blood transfusion effect in man and mice. Kathryn Wood's group[26] has investigated the blood transfusion effect in detail in rodent models and found that combining CD4 antibody with donor specific blood transfusion leads to the generation of donor specific regulatory CD4 T cells with some of the features described above.

A two-step strategy targeting common minors offers many potential benefits. Regulatory T cells may be initiated at a time of relative clinical stability, and then maintained and reinforced by repeated administration at later time intervals. In addition to improved long-term graft acceptance, it may be possible to reduce the level of immunosuppression at the time of transplantation thereby improving patient morbidity and mortality. It has been suggested that indirect recognition of transplant antigens might be the driving force for chronic rejection.[27] Initiating tolerance to minors presented through the indirect route may yet have beneficial effects on chronic rejection which now poses the major long-term problem in transplant immunology.

One may ask whether such strategies can be realistically applied to human transplantation? The long-recognised and well-documented effect of random blood transfusion on kidney graft survival suggests that there is a good chance these therapeutic strategies can be applied. One caveat is that attention now needs to be focussed on the molecular identity of MHC class II restricted minors. We will also need to know whether the minors that are immunodominant for rejection are the same as those immunodominant for suppression.

Acknowledgments
This work was supported by grants from the Medical Research Council.

References
1. Cobbold SP, Waldmann H. Skin allograft rejection by L3/T4⁺ and Lyt-2⁺ T cell subsets. Transplantation 1986: 41:634-639.
2. Benjamin RJ, Waldmann H. Induction of tolerance by monoclonal antibody therapy. Nature 1986; 320:449-451.
3. Benjamin RJ, Cobbold SP, Clark MR et al. Tolerance of rat monoclonal antibodies: Implications for serotherapy. J Exp Med 1986; 163:1539-1552.
4. Gutstein N L, Seaman WE, Scott JH et al. Induction of immune tolerance by administration of monoclonal antibody to L3T4. JImmunol 1986; 137:1127-1132.
5. Carteron NL, Wofsy D, Seaman WE. Induction of immune tolerance during administration of monoclonal antibody to L3T4 does not depend on depletion of L3T4+ cells. JImmunol 1988; 140:713-716.
6. Carteron NL, Schimenti CL, Wofsy D. Treatment of murine lupus with F(ab')₂ fragments of monoclonal antibody to L3T4. Suppression of utoimmunity does not depend on T helper cell depletion. J Immunol 1989; 142:1470-1475.
7. Qin S, Shixin Q, Cobbold SP, Tighe H et al. CD4 monoclonal antibody pairs for immunosuppression and tolerance induction. Eur J Immunol 1987; 17:1159-1165
8. Benjamin RJ, Shixin Q, Wise MP et al. Mechanisms of monoclonal antibody-facilitated tolerance induction: A possible role for the CD4 (L3T4) and CD11a (LFA-1) molecules in self-non-self discrimination. Eur J Immunol 1988; 18:1079-1088.
9. Qin S, Wise M, Cobbold S et al. Induction of tolerance in peripheral T cells with monoclonal antibodies. Eur J Immunol 1990; 20:2737-2745.
10. Qin S, Cobbold S, Benjamin R et al. Induction of classical transplantation tolerance in the adult. J Exp Med 1989; 169:779-794.
11. Wise M, Benjamin R, Qin S et al. Tolerance induction in the peripheral immune system. Molecular mechanisms of immunological self-recognition. Vogel H, Alt F, eds. Academic Press, 1992: 149-155.
12. Scully R, Qin S, Cobbold SP et al. Mechanisms in CD4 antibody-mediated transplantation tolerance: kinetics of induction, antigen dependency and role of regulatory T cells. Eur J Immunol 1994; 24:2383-2392.
13. Davies JD, Leong LYW, Mellor A et al. T-cell suppression in transplantation tolerance through linked recognition. J Immunol 1996; 156:3602-3607.
14. Chen ZK, Cobbold SP, Waldmann H et al. Amplification of natural regulatory immune mechanisms for transplantation tolerance. Transplantation 1996; 62:1200-1206.
15. Qin S, Cobbold SP, Pope H et al. Infectious transplantation tolerance. Science 1993; 259: 974-977.
16. Davies JD, Martin G, Phillips J et al. T-cell regulation in adult transplantation tolerance. J Immunology 1996; 157:529-533.
17. Bemelman F, Honey K, Adams L et al. Bone marrow transplantation induces either clonal deletion or infectious tolerance depending on the dose. J Immunol 1998; 160:2645-2648.
18. Marshall SE, Cobbold SP, Davies et al. Tolerance and suppression in a primed immune system. Transplantation 1996; 62:1614-1621.
19. Waldmann H, Cobbold SP. How may immunosuppression lead to tolerance? The war analogy. In: Banchereau J, Dodet B, Schwartz R et al, eds. Immune Tolerance. Paris: Elsevier, 1996: 221-227.

20. Cobbold SP, Adams E, Marshall SE et al. Mechanisms of peripheral tolerance and suppression induced by monoclonal antibodies to CD4 and CD8. Immunol Rev 1995; 148:1-29.
21. Plain KM, Fava L, Spinelli A et al. Induction of tolerance with non-depleting anti-CD4 monoclonal antibodies is associated with downregulation of Th2 cytokines. Transplantation 1997; 64:1559-67.
22. Waldmann H, Qin S, Cobbold S. Monoclonal antibodies as agents to reinduce tolerance in utoimmunity. J Autoimmunol 1992; 5 Suppl A:93-102.
23. Auchincloss H, Lee R, Shea S et al. The role of "indirect' recognition in initiating rejection of skin grafts from MHC class II deficient mice. Proc Natl Acad Sci USA 1993; 90:3373-3373.
24. Lechler RI and Batchelor JR. Immunogenicity of retransplanted rat kidney allografts. J Exp Med 1982; 156:1835-1841
25. Campos L, Naji A, Deli BC et al. Survival of MHC deficient mouse heterotopic cardiac allografts. Transplantation 1995; 59:187-191.
26. Saitovitch D, Bushell A, Mabbs DW et al. Kinetics of induction of transplantation tolerance with a nondepleting anti-CD4 antibody and donor specific transfusion before transplantation. Transplantation 1996; 61:1642-1647.
27. Vella JP, Vos L, Carpenter CB et al. Role of indirect allorecognition in experimental late acute rejection. Transplantation 1997. 64:1823-1828.

CHAPTER 15

Skin-Specific Minor Histocompatibility Antigens: A Critical Appraisal

David Steinmuller

Discovery of Putative Skin-Specific Histocompatibility Antigens

In 1962 Silverman and Chin[1] reported that some radiation chimeras they had constructed rejected skin grafts from the bone marrow cell (BMC) donor without losing BM chimerism. The model was a xenogeneic one, lethally irradiated C57BL mice restored with Sprague-Dawley rat BMC, and the authors were not the first to report that chimeras do not always accept donor strain skin grafts; in 1959 Billingham and Brent[2] had reported that strain A mice given (C57BL x CBA)F₁ hybrid spleen cells (SC) at birth became stable BM chimeras that accepted CBA but not C57BL skin grafts, a phenomenon they referred to as "split tolerance." However Silverman and Chin were the first to propose an explanation for the effect based on skin-specific histocompatibility (H) antigens. If the hematopoietic system of the BM chimera developed in the absence of donor skin, the chimeras should not be tolerant to donor skin-specific alloantigens; donor skin grafts therefore would be viewed as foreign and rejected. Based on "the long time required for completion of the rejection" (an average of 20-61 days in different experiments) Silverman and Chin likened the skin-specific reactions they observed to reactions against minor H antigens like H3 and HY.[1]

In 1968 Boyse and Old[3] also proposed an explanation for "the loss of skin allograft tolerance by chimeras" based on skin-specific "Sk" antigens; the term later was changed to "Skn" to avoid a nomenclature conflict with the scaly skin mouse mutation, Sk.[4] Boyse et al[5,6] found histogenetic substantiation for the Skn hypothesis: stable BM chimeras constructed by lethally irradiating C57BL/6 (B6) mice and restoring them with (B6 x A)F₁ (B6A) hybrid SC or BMC always accepted B6 skin grafts but consistently rejected both B6A and A skin grafts in 2-4 weeks. However, rejection was considerably delayed if the skin grafts were transplanted immediately after BM restoration rather than some weeks later, and rejection was prevented altogether if the hosts were given an injection of A strain epidermal cells (EC) at the time of BM restoration and 10 weeks later. The interpretation was that loss of self-tolerance to Skn antigens was averted if the hosts were provided with a constant source of donor strain skin cells.

Silvers et al[7] drew the same conclusion from a study of the tolerogenicity of neonatal skin grafts, a phenomenon investigated more recently by Markees et al.[8] BM chimerism was induced in B6 mice by irradiating them at birth and giving them B6A lymph node cells (LNC).[7] As adults the B6A→B6 chimeric hosts rejected adult A strain skin grafts but they

Minor Histocompatibility Antigens: From the Laboratory to the Clinic, edited by Derry Roopenian and Elizabeth Simpson. ©2000 Landes Bioscience.

usually accepted neonatal A skin grafts and most also accepted subsequent grafts of adult A skin. However, if the neonatal grafts were excised before the adult grafts were transplanted, the adult grafts were not accepted, apparently because the source of tolerance-conferring Skn alloantigens had been removed. The rejection of BMC-donor skin grafts by stable BM chimeras has been observed not only in rodents[9] but also in dogs,[10] pigs,[11] humans[12,13] and cattle.[14,15] It is not widely appreciated that synchorial dizygotic cattle twins of the type that awakened Medawar's group[16] to the discovery of acquired tolerance eventually reject each others skin grafts even though they remain lifelong BM chimeras.[14,15] Thus skin-specific H antigens might explain donor type skin graft rejection by stable BM chimeras in a variety of mammalian species.

Antibodies directed against Skn alloantigens have been detected. Scheid et al[17] found that the serum of B6A→B6 radiation chimeras that rejected strain A skin grafts showed complement (C)-dependent cytotoxicity towards strain A EC but not towards B6 EC or a variety of lymphoid cells from either strain. Absorption tests confirmed the results of the direct tests. The cytotoxicity towards A EC was reduced to background levels by absorption with these cells but not with B6 EC or strain A LNC. The cytotoxicity was also absorbed by A but not B6 brain, and the presence of Skn alloantigens on brain cells was confirmed by showing that cells of an A strain neuroblastoma were about as sensitive to lysis as A EC. Thus, A strain skin grafts evoked antibodies by B6A→B6 chimeras that were both tissue- and allospecific. Serum of B6A→A chimeras given B6 skin grafts was cytotoxic to B6 EC but not to A EC, thus defining antigenic products of alternate *Skn* alleles, *Skn^a* and *Skn^b*, in strains A and B6 respectively. Linkage of serologically detected Skn and H2 was excluded because typing tests showed no correlation between the respective antigen phenotypes.[17,18]

The fact that chimeras reject H2-compatible BM-donor skin grafts[19,20] and the relatively slow tempo of rejection induced by Skn antigens, typically 20-100 days, qualifies Skn alloantigens as MHC-restricted "minor" H antigens. Moreover, as with HY transplantation antigens[21] both the immunogenicity of and histogenic response to Skn antigens proved to be under H2 control. Fleming and Silvers[22] found that both donor and host H2 haplotypes influenced the percentage of skin grafts rejected by BM chimeras. For example 88% of A (*H2^a*) versus only 5% of A.BY (*H2^b*) skin grafts were rejected by B10 x A/B10 (*H2$^{b,a}/^b$*) chimeras and only 5% of B10 x A/A (*H2$^{b,a}/H2^a$*) versus 78% of B10 x A/A.BY (*H2$^{b,a}/H2^b$*) chimeras rejected B10 (*H2^b*) skin grafts.

The Number and Mapping of *Skn* Loci

Several studies have investigated the genetics of *Skn* loci.[9] Wachtel et al[23] used backcross mice as hosts for making radiation chimeras that were challenged with Skn-incompatible skin grafts. Eighty-four of 119 B6A-to-(B6A x B6) backcross chimeras (71%) rejected A strain grafts. These results suggested that there are two segregating Skn loci because on this assumption 75% of the grafts would be rejected. Graft rejection times fell into relatively fast (25-30 day), intermediate, and slow (70-80 day) groups suggesting differences in the strength of antigens determined by *Skn* alleles. Rees et al[24] mapped the serologically-defined Skn2.1 alloantigen using monoclonal antibody (MoAb) to type a set of recombinant-inbred (RI) mouse strains, and determined that *Skn2* is linked to the myeloblastosis oncogene, *Myb*. The chromosomal localization was confirmed by typing appropriate backcross segregants, placing *Skn2* 11 ± 6 map units distal to *Myb*.

The many non-*H2* congenic strains selected by Bailey on the basis of skin graft rejection (see Chapter 2) afforded the opportunity to determine whether these congenic strains could differ by *Skn* genes. Steinmuller et al[25] screened 35 of Bailey's bilineal congenic strains by immunizing inbred-partner strain hosts with congenic-strain LNC cells before challenging them with a congenic-strain skin graft. In all cases the LNC either immunized or tolerized the hosts, indicating that the antigens are expressed on lymphoid cells and thus the H differences were not skin-specific. However, Harrison and Mobraaten[20] later reported that (WBxB6)F$_1$-*W/W^v* anemic mutant mice cured with congenic bone marrow from Bailey's B6.C-H24^c (H24) strain rejected H24-disparate skin grafts with a mean survival time (MST) of about 70 days. The discrepancy between the two reports might hinge on Harrison and Mobraaten's use of Bailey's more sensitive tail skin grafting technique. Thus H24, a locus that maps to proximal mouse chromosome 7, appears to encode a Skn antigen.

Problems with Skn Antigens

Skn antigens have been detected by both histogenic and serological methods. Are the two related? The use of C dependent cytotoxicity assays with EC targets is complicated by the fact that normal rodent,[26,27] rabbit[26] and human[28] sera contain natural antibodies that are highly toxic to autologous or syngeneic EC. For example Flaherty and Bennet[27] found that the sera of "about 50% of the mice of all strains tested" contained EC cytotoxins and the levels increased when the skin was abraded. Skin grafting chimeras to raise anti-Skn antibodies certainly abrades the skin, and Scheid et al[17] had to repeatedly absorb selected rabbit sera with mouse EC in the presence of EDTA to obtain a source of C that itself was not toxic to mouse EC targets. Goldberg et al[18,29] avoided the problems of C-dependent cytotoxicity by using immunofluorescence and flow cytometry to characterize Skn antisera and develop Skn MoAbs, but they still had to deal with small quantitative rather than qualitative differences in immunofluorescence of allogeneic and syngeneic EC targets[29] Fluorescence histograms of EC treated with a monoclonal antibody (SK4) reactive with the Skn1.1 antigen showed more impressive shifts with EC from putative Skn1.1$^+$ and Skn1.1$^-$ strains.[18] However the histograms suggested only low levels of expression on few target cells and data controlling for nonspecific binding of the primary and secondary reagents were not presented. Jackman et al[30] reported qualitative differences in immunohistochemical staining of A versus B6 strain epidermis with MoAbs 22.1 and C4 to Skn^a antigens and Rees et al[31] used C4 with an immunoblotting technique to identify a 95-kd protein Skn2.1 antigen in mouse skin and brain lysates. The localization was predominantly cytoplasmic as shown by immunoperoxidase staining on frozen sections of cerebral cortex, cerebellum and hippocampus. In contrast, using a sensitive cellular radioimmunoassay (cRIA) Hadley and Steinmuller (unpublished data) did not find any differences in the binding of sera from B6A→B6 chimeras that rejected A strain skin grafts to A and B6 EC. The difficulties in defining the specificity of Skn antibodies suggest that they might be tissue-specific autoantibodies that demonstrate some degree of allospecificity, a characteristic of human stratum corneum autoantibodies for example.[32]

The presence of serologically defined Skn antigens does not always correlate with skin graft rejection. Not all B6A→A chimeras reject B6 skin grafts, and even those chimeras that accept B6 skin have good titers of cytotoxic antibody to B6 EC.[17] This has its counterpart with the serologically detected male (SDM) HY antigen[21] where females of some strains fail to reject male skin grafts but still produce HY antibodies.[33] This raises the question of whether Skn antibodies play a role in skin graft rejection, usually thought to be a T-cell mediated process.

Attempts to define Skn antigens with T cells are inconclusive. Gillette et al[34] reported that lymphocytes from B6A→B6 chimeras that rejected several A strain skin grafts proliferated in mixed leukocyte culture to A EC but not to B6A LNC cells. However, the response of the chimeras was much weaker than that of normal B6 mice, and because F$_1$ rather than parent-strain cells were used as irradiated LNC stimulators, a hybrid dilution effect might have contributed to the failure to respond to B6A cells. Moreover, as B6 EC were not included as syngeneic control stimulators, the allospecificity of the putative T-cell recognition of Skn antigens was not clearly established. Burlingham and Steinmuller (unpublished data) failed to detect Skn-specific cytotoxic T cells (CTL) in B6A→B6 radiation chimeras even after the rejection of A strain skin grafts; the only CTL they detected were equally if not more cytotoxic to A LNC than EC. Jackman (personal communication) also found that the Skn-specific CTL she tried to generate from B6A→B6 chimeras reacted with both EC and lymphocyte A strain targets. These results suggest that rather than being B6A donor cells responsive to Skn antigens, the CTL detected in these assays probably originated from residual alloreactive B6 host cells. Van Bekkum and Roodenburg[35] reported that SC from (CBA x BL)F$_1$-to-BL chimeras that rejected CBA or F$_1$ skin grafts induced lethal graft-versus-host disease (GVHD) in secondary F$_1$ hybrid hosts. Fleming et al[36] induced fatal GVHD in B6AF$_1$ hosts with SC from B6A→B6 chimeras that rejected A strain skin grafts and they noted that "the severity of the reactions observed in these F$_1$ hybrid mice was similar to that observed in F$_1$ mice that received H2 compatible B6 cells." Suman et al[37] used the T6 chromosome marker to demonstrate small numbers of residual host cells in persistent radiation chimeras that rejected bone marrow-donor skin grafts. They concluded that residual host rather than BMC-donor cells mediated rejection of the skin grafts. Dr. Peter Démant (personal communication) made a similar observation.

Jackman et al[38] addressed the crucial question of which cells mediate Skn immunity in an adoptive transfer experiment. B6A hosts given an intravenous injection of SC from B6A→B6 radiation chimeras developed severe GVHD-like skin lesions if:

1. the chimera donors were primed by the rejection of an A strain skin graft,
2. the recipients' skin was stimulated by shaving—the lesions were confined to the shaved skin areas, and
3. the recipients were pretreated with sublethal doses of cytoxan (CTX) or whole-body irradiation.

SC from female donors primed to HY by the rejection of a syngeneic male skin graft did not induce lesions in CTX-treated female B6A hosts. Most importantly, pretreatment of the inocula with antiserum directly against strain A H2 antigens abolished their capacity to damage the skin, indicating that the lesions were induced by B6A donor not residual B6 host cells. The authors compared these results to CTX and cyclosporine-induced syngeneic GVHD in the rat,[39] which appears to be mediated by unregulated autoreactive T cells. Jackman et al[38] also reported that adoptively transferred SC from B6A chimeras that had rejected a strain A skin graft induced the rejection of strain A but not B6 skin grafts by B6A hosts: all B6A hosts pretreated with CTX and 2 of 5 non-pretreated B6A hosts rejected A strain skin grafts in 18-37 days. However, given the importance of proving that rejection of the A strain skin grafts was mediated by Skn-reactive B6A cells from the chimeras and not by residual H2-reactive B6 host cells, it is unfortunate that Jackman et al[38] did not test the effect of pretreating the inocula with anti-H2^a serum as they did in the skin lesion experiment. Moreover they did not separate or characterize the effector cells in either experiment in terms of surface antigens or subset markers.

As for the histogenetic evidence of Skn antigens, when Boyse et al[6] putatively maintained self-tolerance of A strain Skn antigens in B6A→B6 chimeras with injections of A EC, they did not control the experiment with injections of equivalent numbers of A LNC that might have induced a more complete tolerance of conventional H-antigen differences between the A and B6 strains. Moreover when Silvers et al[7] broke tolerance by excising neonatal A strain skin grafts on B6A→B6 chimeras neither the BM origin[40] nor the tolerogenicity[41] of epidermal dendritic cells was known. Thus they did not consider the possibility that "passenger" BM-derived cells carried in the neonatal skin grafts, which by definition do not express Skn antigens, were the source of tolerance induction. In summary, it is not clear whether Skn antigens are alloantigens or autoantigens, and whether they evoke T-cell responses, especially in view of the aforementioned failures to generate Skn-specific CTL.

Detection and Analysis of Epidermal H Antigens

Steinmuller et al[42] identified a new antigen in mouse skin when they found that immunizing C3H/He mice with H2^k-compatible AKR or CBA EC generates H2-restricted CTL that preferentially lyse donor EC as opposed to LNC targets. The CTL are allospecific because they do not lyse syngeneic EC above background level. Genetic mapping and backcross studies[43] revealed that the restricting element is a product of the *H2K* region and that the target antigen, epidermal alloantigen-1 (Epa-1, now Epa1) is determined by a single Mendelian gene not linked to *Skn* loci. Although Epa1-specific CTL lyse EC of different Epa1$^+$ strains at different levels, assays with an H2K^k-specific MoAb indicated that this reflects quantitative differences in expression of the H2K-restricting element rather than polymorphism of the *Epa1* gene.[44] Moreover the fact that Epa1-specific CTL are generated in one direction only in reciprocal tests with several *H2^k*-compatible strain combinations[42] indicates that *Epa1* apparently was revealed by a loss mutation. Tests with appropriate combinations of *H2* congenic lines as donors and hosts showed that the ability to generate Epa1-specific CTL is under *H2* gene control with *K/D* rather than I-region products serving as immune-response genes.[45] Cold-target inhibition and antigen-driven suicide assays[46] revealed the presence of two CTL populations in bulk cultures generated by immunization with H2-compatible EC, a small one reactive with antigens shared by EC and LNC that accounts for the low level of lysis of LNC targets by bulk-culture CTL and a large population reactive with EC exclusively. The two populations were separated by limiting dilution and several Epa1-specific CD8$^+$ CTL clones were established that had the same lytic and *H2* restriction specificities as bulk- culture Epa1 CTL.[47] Subsequently H2-restricted Epa1 CTL were cloned from EC-impregnated sponge-matrix allografts in vivo[48] demonstrating that the CTL were graft-infiltrating effector cells.

As for tissue specificity, although Epa1 is not a CTL target cell determinant on lymphoid cells, it is expressed as a weak immunogen by at least some of these cells. For example, in limiting dilution analyses of Epa1-specific CTL precursors, CBA SC immuned C3H/He hosts for subsequent boosting with CBA EC, though much less efficiently than CBA EC.[46] Epa1 CTL also were generated when SC from C3H/He hosts immunized with CBA EC were boosted in vitro with CBA SC[46] showing that CBA SC can substitute to some extent for CBA EC at either the immunization or boosting stage of Epa1 CTL generation. In addition, BM chimeras were used as a source of stimulator cells to test the influence of BM-derived cells in the skin on Epa1-specific CTL generation.[49] By making reciprocal radiation chimeras between Epa1$^+$ and Epa1$^-$ strains, EC suspensions were available in which the keratinocytes and BM-derived cells in the epidermis came from strains of reciprocal Epa1 phenotypes. EC suspensions from Epa1$^-$ hosts with long-established Epa1$^+$-strain BM showed weak but significant immunizing and boosting ability, proving that Epa1 is expressed to some extent as an immunogen on BM-derived EC such as Langerhans cells and other skin dendritic cells. In addition although fresh peritoneal macrophages are resistant to lysis by Epa1 CTL clones they become susceptible after 12-24 hours in vitro.[50] However fresh macrophages are lysed if they came from mice treated with the macrophage-activating agents like Con A and BCG but not from mice treated with the sterile inflammatory agents like peptone broth and thioglycollate. The correlation between Epa1 expression and macrophage-activation suggests that Epa1 might be an allospecific marker for activated macrophages as well as an inducible H antigen in vitro. Moreover the presence of small numbers of Epa1$^+$ activated macrophages in the spleen helps to explain the weak though significant ability of SC to immunize and boost the generation of Epa1-specific CTL.[46] Epa1-specific CTL also lyse cultured EC[50] indicating that the preferential lysis of fresh EC is not an artifact of the trypsinization required to prepare single-cell suspensions from the epidermis. The CTL also lyse fibroblasts cultured from abdominal fascia as well as from skin[51] suggesting that Epa1 might be expressed on fibroblasts.

When cloned Epa CTL are injected intradermally into Epa1$^+$ hosts as few as 1 x 10^6 evoke gross skin lesions marked by edema, infiltration of the dermis and epidermis and full-thickness necrosis of the skin by three days.[52,53] In contrast there is no skin necrosis in Epa1$^-$ hosts and the CTL remain at the injection site indicating that their migration into and destruction of the epidermis is triggered by exposure to Epa1. Moreover, the ability of Epa1 CTL clones to attack and destroy skin is just as H2-restricted as their lytic activity in vitro; skin lesions develop only if the hosts express both Epa1 and the required *H2K* restriction element.[52,53] Furthermore unlike the adoptive transfer experiments in the Skn system[38] there is no requirement to treat hosts with immunosuppressive agents to evoke skin lesions with Epa CTL.

To determine if Epa1 is an H antigen, (C3H/He x CBA)F$_1$ hybrids were backcrossed to the C3H parent strain. The resulting offspring then were typed for Epa1 and the survival of CBA skin grafts on Epa1 compatible and incompatible hosts were compared.[54] Even though C3H/He and CBA mice are disparate for numerous minor H antigens in addition to Epa1 merely matching for Epa1 resulted in a striking incidence of prolonged graft survival. Subsequently an Epa1$^+$ congenic strain, C3H. Epa1 was developed on the C3H/He background and the survival times of skin and heart transplants exchanged between the congenic strain and its inbred partner were compared.[55] All C3H females and males rejected first-set Epa1 skin grafts with MSTs of 20 and 31 days respectively with survival times ranging from 12-71 days, the more vigorous response of female hosts being characteristic of non-H2 skin graft rejection in general.[56] However there was a very strong factor of immunization; second-set Epa1 skin grafts transplanted two months later were all rejected within 10 days with MSTs of six days in both sexes. As for the Epa heart allografts, most were rejected by C3H hosts in chronic fashion with some grafts lingering on for more than 100 days.[55] Moreover in contrast to the markedly accelerated rejection of second-set skin grafts, Epa$^+$ hearts transplanted to C3H hosts sensitized by the rejection of an Epa$^+$ skin allograft also were chronically rejected. However the fact that C3H mice reject Epa1 congenic hearts at all indicates that Epa1 must be expressed on heart tissue. In fact Epa1 might be expressed on a variety of other parenchymal and reticular tissues but this is difficult to determine because of the inability to raise Epa1 antibodies[57] that could be used for in situ immunofluorescence or immunochemistry or for the development of molecular probes for tests of *Epa1* gene expression in various tissues.

The tests with the Epa congenic strain also is consistent with the interpretation that *Epa1* was the manifestation of a loss mutation, a hypothesis previously based on the finding that Epa1-specific CTL could be generated in one direction only in several reciprocal Epa1 disparate mouse strain combinations.[42] In contrast to the rejection of Epa-congenic skin and heart grafts by C3H hosts, in the reciprocal direction all first- and second-set C3H skin and heart grafts were accepted permanently by Epa1$^+$ congenic hosts.[55] Thus mutant (C3H/He) hosts reject wild-type (C3H.Epa1) hosts and accept mutant (C3H/He) grafts. Moreover when the six different extant C3H mouse strains were typed for Epa1[58] it was found that the three derived from Strong's original line, C3H/St, C3H/Bi and C3H/Mai are Epa1$^+$ whereas the three derived from Andervont's offshoot, C3H/An, C3H/He and C3HeB/Fe are Epa1$^-$. Thus fixation of the Epa1 loss mutation must have occurred soon after the Strong and Andervont C3H lines diverged in 1931.

It is particularly interesting that the loss mutation at the *Epa1* locus enables C3H/He mice to respond to EC of other mammalian species. C3H mice were immunized with rat, human or chicken EC and host SC then were boosted with CBA EC in an attempt to generate CTL that would preferentially lyse CBA EC targets.[59] When compared with control CTL generated by immunizing with CBA EC, EC from 15 different rat strains immunized just as well, and EC from 27 different humans on average immunized 40% as well as CBA EC. Neither human peripheral blood leukocytes nor chicken EC immunized C3H mice for Epa1 CTL indicating the presence of tissue-restricted alloantigens in rat and human but not chicken epidermis that cross-react with Epa1 at the immunizing stage at least.

Evidence for Additional Epa Antigens

Sakai et al[60] found that allogeneic EC stimulated SC in RT1-identical, mixed lymphocyte-reaction (MLR) negative rat strain combinations and that the magnitude of the EC stimulation and skin allograft survival times were well correlated. Hirschberg and Thorsby[61] used a hot pulse of tritiated thymidine to specifically eliminate responder lymphoid cells proliferating in one-way human MLRs; the remaining lymphoid cells no longer responded upon restimulation with donor lymphoid cells but they did respond to donor EC. These experiments suggest the presence of T-cell-activating minor alloantigens in rat and human skin that might be Epa-like antigens.

Although *Epa1* is not obviously polymorphic there is suggestive evidence of other *Epa* loci. The *H2*-identical B10.BR and CBA strains both are Epa1+and EC from both strains stimulate Epa1-specific CTL in C3H/He mice. However B10.BR hosts can be immunized with CBA EC to generate CTL that preferentially lyse CBA EC. Moreover, cold-target inhibition assays in this *Epa1* identical strain combination indicated the presence of EC-specific CTL that therefore would be directed against the product of a second Epa locus provisionally designated *Epa2* (Steinmuller and Tyler, unpublished data). Similarly immunization with EC in two other *H2* identical strain combinations, the Epa1+ B10.S-to-SJL and the Epa1‾ A/J-to-C3H.A combinations evokes CTL that preferentially lyse donor EC targets suggesting the presence of still other *Epa* genes. In all three strain combinations EC-specific CTL are generated in one direction only again indicating the presence of silent alleles. Thus *Epa1* might be representative of a class of nonpolymorphic genes whose antigenic products evoke CTL that may attack skin and other allografts if the hosts happen to carry the appropriate loss mutations. However like Skn antigens the molecular nature of Epa antigens as well as the normal function of *Epa* genes still are undetermined.

References

1. Silverman MS, Chin PH. Differences in the rejection of skin transplants and blood cells of donor marrow origin by radiation-induced chimeras. Ann NY Acad Sci 1962; 99:542-549.
2. Billingham RE, Brent L. Quantitative studies on tissue transplantation immunity. IV. Induction of tolerance in newborn mice and studies on the phenomenon of runt disease. Philos Trans R Soc Lond Biol 1959; 42:439-477.
3. Boyse EA, Old LJ. Loss of allograft tolerance by chimeras. Transplantation 1968; 6:619.
4. Lyon MF, Rastan S, Brown SDM. Genetic variants and strains of the laboratory mouse. 3rd ed. Oxford: Oxford University Press, 1996:715.
5. Boyse EA, Lance EM, Carswell EA et al. Rejection of skin allografts by radiation chimaeras: Selective gene action in the specification of cell surface structure. Nature 1970; 227:901-903.
6. Boyse EA, Carswell EA, Scheid MP et al. Tolerance of Sk incompatible skin grafts. Nature 1973; 244:441-442.
7. Silvers WK, Wachtel SS, Poole TW. The behavior of skin grafts incompatible with respect to skin alloantigens on mice rendered tolerance at birth with lymphoid cells. J Exp Med 1976; 143:1317-1326.

8. Markees TG, De Fazio SR, Gozzo JJ. Prolonged adult skin allograft survival as a result of cotransplantation with neonatal tissue. Transplantation 1992; 54:955-958.

9. Steinmuller D, Wachtel SS. Transplantation biology and immunogenetics of skin-specific (Sk) alloantigens. Transplant Proc 1980; 12, Suppl 1:100-106.

10. Rapaport FT, Lawrence HS, Bachvaroff R et al. Histocompatibility studies in a closely bred colony of dogs. V. Mechanisms of cellular adaption in long-term DL-A identical radiation chimeras. J Exp Med 1975; 142:120-138.

11. Smith CV, Nakajima K, Mixon A et al. Successful induction of long-term specific tolerance to fully allogeneic renal allografts in miniature swine. Transplantation 1992; 53:438-444.

12. Woodruff MFA, Lennox B. Reciprocal skin grafts in a pair of twins showing blood chimerism. Lancet 1959; 2:476-478.

13. Beilby JOW, Cade IS, Jelliffe AM et al. Prolonged survival of a bone-marrow graft resulting in a blood-group chimera. Br Med J 1960; 2:96-99.

14. Stone WH, Cragle RG, Swanson EW et al. Skin grafts: Delayed rejection between pairs of cattle twins showing erythrocyte chimerism. Science 1965; 148:1335-1336.

15. Emery D, McCullough P. Immunological reactivity between chimeric cattle twins. I. Homograft reaction. Transplantation 1980; 29:4-9.

16. Billingham RE, Lampkin GH, Medawar PB et al. Tolerance to homografts, twin diagnosis, and the freemartin condition in cattle. Heredity 1952; 6:201-212.

17. Scheid M, Boyse EA, Carswell EA et al. Serologically demonstrable alloantigens of mouse epidermal cells. J Exp Med 1972; 35:938-955.

18. Goldberg EH, Goble R, Jackman SH. Monoclonal antibodies defining skin-selective alloantigens, Skn. Immunogenetics 1990; 1:393395.

19. Warner NL, Herzenberg LA, Cole LJ et al. Dissociation of skin homograft tolerance and donor type gamma globulin synthesis in allogeneic mouse radiation chimeras. Nature 1965; 205:1077-1079

20. Harrison DE, Mobraaten LE: Skin graft rejection in mice repopulated with marrow of the skin donor type: An Skn gene in a congenic line. Immunogenetics 1984; 19:503-509.

21. Simpson E, Scott D, Chandler P. The male-specific histocompatibility antigen, HY: A history of transplantation immune response genes, sex determination and expression cloning. Annu Rev Immunol 1997; 15:39-61.

22. Fleming HI, Silvers WK. An immunogenetic analysis of Skn antigens in mice. Immunogenetics 1981; 14:517-526.

23. Wachtel SS, Thaler HT, Boyse EA. A second system of alloantigens expressed selectively on epidermal cells of the mouse. Immunogenetics 1977; 5:17-23.

24. Rees D, Nesbitt MN, Goldberg EH. Skn 2 is linked to Myb on chromosome 10 of the mouse. Immunogenetics 1994; 9:363-366.

25. Steinmuller D, Marcus JL, Bailey DW. The screening of non-H2 congenic lines for Skn loci. Immunogenetics 1978; 7:239-245.

26. Terasaki PI, Chamberlain CC: Destruction of epidermal cells in vitro by autologous serum from normal animals. J Exp Med 1962; 115:439-452.

27. Flaherty L, Bennett D. The unusual fate of second skin grafts across an H(Tla) barrier. Transplantation 1973; 16:682-684.

28. Ackermann-Schopff C, Ackermann R, Terasaki PI et al. Natural and acquired epidermal autoantibodies in man. J Immunol 1974; 112:2063-2067.

29. Jackman SH, Goldberg EH. Demonstration of Skn antigens on mouse epidermal cells by immunofluorescence and flow cytometry. J Invest Dermatol 1987; 88:574-576.

30. Jackman SH, De Pirro ES, Goldberg EH. Immunohistochemical identification of Skn antigens in mouse epidermis. J Invest Dermatol 1989; 93:46-49.

31. Rees D, Carey F, Goldberg EH. Skn antigens: Identification of a 95-kilodalton protein in mouse neural tissue. Reg Immunol 1993; 5:94-99.

32. Dabski K, Beutner EH. Demonstration of allospecific differences in human stratum corneum autoantibodies by blocking immunofluorescence. Int Arch Allergy Appl Immunol 1985; 77:281-286.

33. Goldberg E, Boyse EA, Scheid M et al. Production of HY antibody by female mice that fail to reject male skin. Nature New Biol 1972; 238:55-56.

34. Gillette RW, Cooper S, Lance EM. The reactivity of murine lymphocytes to epidermal cells. Immunology 1972; 3:769-776.

35. Van Bekkum DW, Roodenburg J. Antidonor immunologic reactions of long-standing radiation chimeras. Transplant Proc 1973; 5:881-886.

36. Fleming HL, Elkins WL, Silvers WK. Investigation of the capacity of skin-specific alloantigens to elicit graft-versus-host reactions in the mouse. Transplant Proc 1981; 13:1223-1225.

37. Suman L, Silobrcic V, Kastelan A. Chimerism in lymph nodes of F₁ into irradiated parental recipient chimeras rejecting skin allografts from the other parent strain. Transplantation 1978; 25:302-304.

38. Jackman SH, Boyse EA, Goldberg EH. Adoptive transfer of skin-selective utoimmunity by Skn alloantigenic disparities. Proc Natl Acad Sci USA 1992; 89:11041-11045.

39. Hess AD, Fischer AC. Immune mechanisms in cyclosporine-induced syngeneic graft-versus-host disease. Transplantation 1989; 48:895-900.

40. Wolff K, Stingl G. The Langerhans cell. J Invest Dermatol 1983; 80, Suppl 6:17s-21s.

41. Steptoe RJ, Thomson AW. Dendritic cells and tolerance induction. Clin Exp Immunol 1996; 105:397-402.

42. Steinmuller D, Tyler JD, David CS. Cell-mediated cytotoxicity to non-MHC alloantigens on mouse epidermal cells. I. H-2 restricted reactions among strains sharing the H2ᵏ haplotype. J Immunol 1981; 126:17471753.

43. Steinmuller D, Tyler JD, David CS. Cell-mediated cytotoxicity to non-MHC alloantigens on mouse epidermal cells. II. Genetic basis of the response of C3H mice. J Immunol 1981; 126:1754-1758.

44. Hadley GA, Snider ME, Steinmuller D. Strain differences in the expression of the Epa1 restricting element. J Immunogenet 1987; 14:149-158.

45. Tyler JD, Burlingham WJ, David CS et al. Cell-mediated cytotoxicity to non-MHC alloantigens on mouse epidermal cells. IV. Influence of the MHC on the tissue specificity of cytotoxic T-lymphocyte responses. Immunogenetics 1982; 16:23-26.

46. Tyler JD, Steinmuller D. Cell-mediated cytotoxicity to non-MHC alloantigens on mouse epidermal cells. III. Epidermal cell-specific cytotoxic T lymphocytes. J Immunol 1981; 126:1759-1763.

47. Tyler JD, Steinmuller D. Establishment of cytolytic T-lymphocyte clones to epidermal alloantigen Epa-1. Transplantation 1982; 4:140-143.

48. Snider ME, Armstrong L, Hudson JL et al. In vitro and in vivo cytotoxicity of T cells cloned from rejecting allografts. Transplantation 1986; 42:171-177.

49. Burlingham WJ, Steinmuller D. Cell-mediated cytotoxicity to nonmajor histocompatibility complex alloantigens on mouse epidermal cells. V. Contribution of bone marrow-derived cells to Epa1 antigen expression. Transplantation 1983; 35:130-135.

50. Burlingham WJ, Snider ME, Tyler JD et al. Lysis of mouse macrophages, fibroblasts and epidermal cells by epidermal alloantigen-specific CTL. Effect of culture and inflammatory agents on Epa1 expression. Cell Immunol 1984; 87:553-565.

51. Burlingham WJ, Tyler JD, Steinmuller D. Cytotoxic T cells specific for epidermal alloantigen, Epa-1, lyse allogeneic fibroblasts and macrophages in addition to epidermal cells. Transplant Proc 1983; 15:292-295.

52. Tyler JD, Steinmuller D, Galli SJ et al. Allospecific graft-versus-host lesions mediated in MHC-restricted fashion by cloned cytolytic T lymphocytes. Transplant Proc 1983; 15:1441-1445.

53. Tyler JD, Galli SJ, Snider ME et al. Cloned Lyt-2⁺ cytolytic T lymphocytes destroy allogeneic tissue in vivo. J Exp Med 1984; 159:234-243.

54. Steinmuller D, Tyler JD, Waddick et al. Epidermal alloantigen and the survival of mouse skin allografts. Transplantation 1982; 33:308-313.

55. Steinmuller D, Wakely E, Landis SK. Evidence that epidermal alloantigen Epa1 is an immunogen for murine heart as well as skin allograft rejection. Transplantation 1991; 51:459-463.
56. Hildemann WH. Components and concepts of antigenic strength. Transplant Rev 1970; 3:5-2.
57. Hadley GA, Steinmuller D. Domination by epidermal cell-reactive autoantibodies in attempts to raise antibodies to Epa1 alloantigen. Transplantation 1988; 45:215-221.
58. Steinmuller D, Tyler JD, Zinsmeister AR. Strain distribution of the new tissue-restricted alloantigen Epa1. Transplant Proc 1985; 7:749-753.
59. Steinmuller D, Tyler JD. Cross-priming reveals similar tissue-restricted CTL-defined alloantigens on mouse, rat and human epidermal cells. Transplant Proc 1983; 15:238-241.
60. Sakai A, Kashiwabara H, Taha M et al. Role of organ-specific antigens in the AgB-compatible graft rejection. Transplant Proc 1977; 9:629-632.
61. Hirschberg H, Thorsby E. Lymphocyte activating alloantigens on human epidermal cells. Tissue Antigens 1975; 6:183-194.

CHAPTER 16

Antigens Encountered During Organ Transplantation

Thalachallour Mohanakumar, Craig R. Smith, Bashoo Naziruddin and Nancy J. Poindexter

Transplantation of bone marrow and solid organs initiates a complex array of immunological responses in the recipient. A perfect match at both major and minor histocompatibility (H) antigens is a very rare occurrence. Therefore, immune system of the recipient is confronted with an array of foreign antigens in the form of MHC-encoded molecules and bound peptides. Study of renal allograft infiltrating lymphocytes (GIL) from biopsies of individuals undergoing acute cellular rejection indicate that the allograft response includes T cells that recognize tissue-specific antigens. In cases in which there are MHC mismatches between the recipients and the donor organ these antigens can be defined as alloantigenic MHC-derived tissue-specific peptides (Table 16.1). When there is a complete HLA match between donor and recipient both conventional and tissue-specific minor H antigens can be recognized. Tissue-specific antigens may also be in involved in autoimmune disease.

Evidence for Tissue-Specific Alloantigens

Organ specific non-MHC antigens of kidneys can evoke both humoral and cellular immune responses. Hart and Fabre[1] have demonstrated the presence of a non-MHC kidney-specific alloantigen that induces a strong anti-tubular basement membrane antibody response. Antibodies that react specifically with donor, but not host tubular basement membrane, appear in the serum of allograft recipient rats, suggesting non-MHC kidney specific alloantigens.[2] Studies on murine interstitial nephritis have also suggested a role for anti-tubular basement membrane antibodies which react with tubules from susceptible strains but not others.[3] Basement membrane antigens may also differ among humans, in that differences in human glomerular basement membrane (GBM) antigens have also been noted in individuals with hereditary nephritis of the Alports type.[4] However, very little is known about the biochemical, functional or molecular properties that give rise to serologically-detected expressed selectively in kidneys.

There are more convincing reports of T-cell detected antigens that are kidney specific. Mashimo et al[5] reported that rat kidney cells stimulate lymphocyte blastogenesis in one of four MHC compatible rat strains. Kidney associated antigens can also induce lymphocyte blastogenesis in dogs.[5-7] Vegt et al[6] showed by absorption studies a population of CTL in dogs which can lyse kidney cells but not autologous phytohemagglutinin (PHA) blasts, indicating organ specific antigens on dog kidney cells. Esquenazi et al[7] have further concluded that two signals are necessary for the canine MLR, one involving kidney associated nominal antigenic epitopes and the other involving class II molecules. Studies in humans have also

Minor Histocompatibility Antigens: From the Laboratory to the Clinic, edited by Derry Roopenian and Elizabeth Simpson. ©2000 Landes Bioscience.

Table 16.1. Antigens recognized by allograft-specific T cells

Antigen	Definition
Tissue-specific MHC alloantigens	Peptides arising from proteins expressed in a tissue-specific manner that are presented by donor but not recipient MHC.
Tissue-specific serologically-defined alloantigens	Tissue-specific antigens that elicit antibody-mediated immunity.
Conventional minor H antigens	Polymorphic peptides that are presented by self-MHC molecules shared by donor and recipient.
Tissue-specific minor H antigens	Minor H antigens expressed in a tissue-specific manner.
MHC presented autoantigens	Self peptides (possibly altered) presented by self MHC.

indicated the presence of kidney specific non-MHC antigens which can elicit a cellular immune response. Manca et al[8] reported that lymphocytes extracted from rejected kidneys specifically lysed kidney fibroblasts but not lymphocytes. Roth et al[9,10] reported that kidney cortical cells stimulated autologous and allogeneic lymphocytes in MLC culture and furthermore correlated reactivity with the in vivo state of the donor cell: there was an enhanced stimulatory capacity of renal cortical cells extracted from kidneys with end stage renal disease and from rejected grafts, both in allogeneic and autologous responses. They concluded that tissue associated determinants contribute to the stimulus observed, but the nature of MHC association remains unclear.

Evidence for the presence of tissue-specific antigens presented by allogeneic MHC molecules that may play a role in allograft rejection comes from situations when multiple organs from the same donor are simultaneously transplanted into a recipient but rejection of only one of the organs is observed. For example, simultaneous kidney-pancreas (SKP) transplantation is a widely accepted treatment for diabetic patients with chronic renal failure. Between October 1987 and December 1994 approximately 3300 SKP transplants were performed in the United States.[11] In an attempt to understand the reasons for better graft survival seen in the pancreas allograft following SKP transplants, one study focused on the frequency of rejection episodes, noting the kidney to undergo a greater frequency of rejection (0.81/patient) compared with the pancreas (0.41/patient).[12] One would expect that if rejection is mediated only by HLA disparities, both organs should have similar graft survival. Recent reports demonstrating the presence of alloreactive CTL specific for tissue-specific antigens expressed on the kidney support the concept that tissue-specific antigens presented in the context of allo-MHC may be involved in allograft immunity.[13]

Another example comes from cardiopulmonary transplantation which is commonly performed for end-stage cardiac and pulmonary failure. Between October 1987 and December 1994 over 400 heart-lung transplants were performed with 1 year graft survival rates of 62.5%.[11] Clinical experience suggested that following cardiopulmonary transplantation, rejection of the lung did not occur in the absence of cardiac rejection. However, rejection can be asynchronous[14] indicating a possible role for tissue-specific allo-MHC antigens. Reactivity to these antigens could explain differences in organ survival in multiple organ transplant recipients.

An experimental model of allograft tolerance also strongly suggests a role for tissue-specific antigens in allograft rejection. Intrathymic injection of donor alloantigenic splenocytes with transient depletion of peripheral lymphocytes (by intraperitoneal injection of antilymphocytic serum) has been shown to induce indefinite donor-specific tolerance to rat cardiac allografts in 88% of recipients.[15] This same protocol does not result in indefinite renal allograft survival, suggesting that tissue-specific antigens not expressed on the injected splenocytes may be expressed on the kidney.

Identification of T Cells Recognizing Tissue-Specific MHC Alloantigens

Kidney Transplantation
Highly polymorphic MHC class I antigens were first identified because of their ability to elicit a strong transplantation rejection response. The finding that T cells recognize MHC plus peptide has added an layer of complexity to the rejection response. T cells, in the case of an allograft, not only can recognize allo-MHC determinants independently of a specific peptide but also allo-MHC plus specific peptides. In cases of an MHC-matched grafts either from living related or cadaver donors, T cells will distinguish between self-MHC with self-peptides and self-MHC with non-self peptides, i.e., minor H peptides. After solid organ allotransplants the recipient immune system could also respond to tissue-specific alloantigens with a humoral response.

In renal transplantation, kidney specific responses have been described at the humoral and cellular level. We and others,[16] have eluted from rejected human kidney allografts antibodies that react specifically with kidney cells and not hemopoietic cells. T lymphocytes that recognize kidney specific allo-MHC antigens have also been identified. Graft infiltrating lymphocytes (GIL) isolated from either nephrectomized kidneys or needle aspirate biopsies of individuals undergoing acute cellular rejection contain donor specific CTL.[17,18] The majority of these T cells are allorestricted, and also recognize donor HLA on lymphoid targets. Using primary kidney epithelial cell lines (KCL) from kidney as well as B lymphoblastoid cells (LCL) of donor, we examined the specificity of GIL populations demonstrating donor specific lysis of KCL and LCL targets (Fig. 16.1). Some of these T-cell lines (NT, YS, JK3, and JL2) appear to be tissue-specific in that they recognize the KCL to a greater extent than the HLA identical LCL. Some of the T-cell lines (LMM, GB, and JG4) recognized both KCL and LCL suggesting these GIL recognize an MHC alloantigen presented by lymphocytes and kidney cells. Through cold target inhibition analysis (Fig. 16.2) we showed that some of these GIL contain tissue-specific T cells recognize donor KCL and not donor LCL. Limit dilution cloning of these populations, using the donor KCL as stimulator cells, has shown that the frequency of tissue-specific T cells in GIL is between 10 and 15% of the total clones isolated.[18,19] These tissue-specific T-cell clones were primarily CD8$^+$ CTL while a small number, isolated from a single patient, were CD4$^+$ CTL clones. The target specificity of CD8$^+$ clones from three separate cloning experiments is presented in Table 16.2. Van der Woude has also shown that kidney specific CTL can be isolated from GIL of renal transplant recipients undergoing acute rejection.[20,21] They identified T-cell lines that recognized both donor KCL and splenocytes as well as T-cell lines specific for either KCL or splenocytes. In addition, in patients with moderate to severe rejection, GIL lines displayed a CD8 independent recognition of KCL whereas recognition of splenocytes from the same organ donor was CD8 dependent. Taken together, these findings suggest that

Table 16.2. CD8⁺ T-cell clones derived from biopsies of renal transplant patients undergoing acute cellular rejection

Biopsy	Source of Biopsy	Stimulator	#CD8$^+$ Clones[a]					n
			I	II	III	IV	V	
JGB	Nephrectomy	donor KCL	4	11	8	1	0	24
LMM	Biopsy	donor KCL	3	8	3	0	2	16
GB	Nephrectomy	donor KCL	2	22	0	1	0	25

[a], CD8$^+$ T cell clones are divided into groups I through V based on their reactivity with donor derived targets.
I = CTL which are KCL specific lysing only the donor KCL
II = CTL which lyse both donor KCL and LCL
III = CTL which lyse only donor LCL
IV = Non-cytolytic clones
V = Non-specific CTL which lyse third party targets as well as donor KCL and LCL

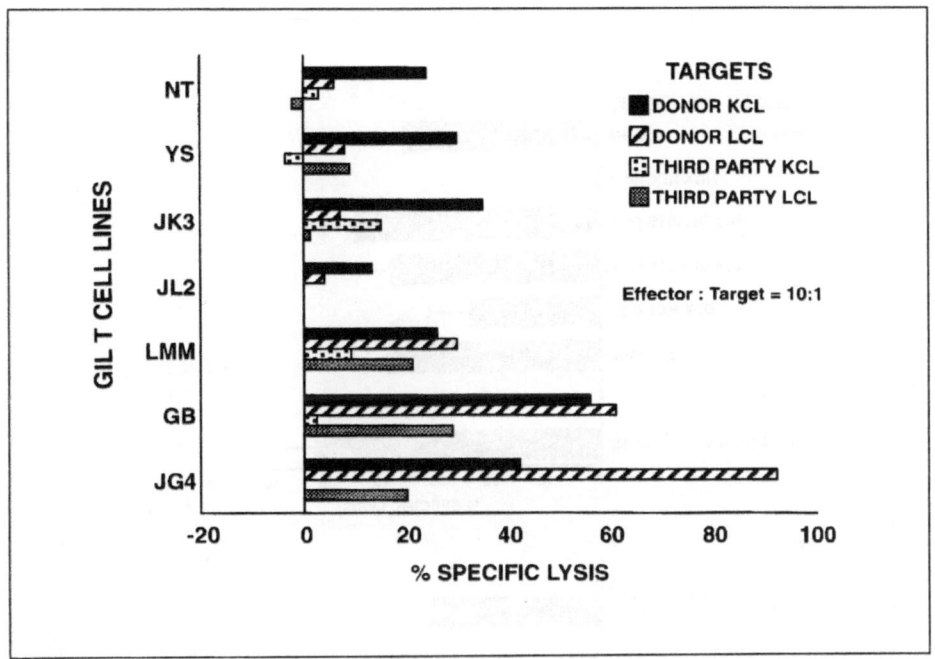

Fig. 16.1. T cells derived from allograft infiltrating lymphocytes (GIL) from biopsies of renal transplant recipients undergoing acute cellular rejection recognized donor antigens on KCL and LCL. GIL were grown in 20 U/ml rIL-2 for 7 to 14 days and then tested for cytolytic activity by standard ^{51}Cr release assays using donor KCL and LCL as targets at an effector to target ratio of 10 to 1. Third party KCL and LCL shared no HLA antigens with donor and, therefore, served as non-specific lysis control targets. Cell lines NT, YS, JK2, and JL2 appeared to be kidney tissue-specific, lysing donor KCL but not the HLA identical donor LCL. While LMM, GB and JG4 were allospecific, not distinguishing between donor HLA on KCL and LCL targets.

tissue-specific antigen recognition mediated by both CD4$^+$ and CD8$^+$ T cells play an important role in kidney allograft rejection.

Heart Transplantation

Tissue-specific antigens presented by allo MHC molecules have also been documented following heart transplantation by examining the activity of cardiac graft infiltrating cells cultured from endomyocardial biopsies.[22] One GIL culture was isolated from the biopsy of a heart transplant patient 203 days post transplant showing a grade 3 rejection, requiring treatment; and the other GIL from a heart transplant patient day 1434 post-transplant showing no rejection. It was shown that these GIL exhibited cytotoxic reactivity specifically directed against allo-MHC/peptide complexes on donor heart endothelial cells.

In Vitro Generated Tissue-Specific Responses

To understand the molecular nature of the tissue-specific response of GIL from renal transplant recipients we have examined in vitro the immunogenicity of parenchymal cells of the human kidney.[23] Tubular epithelial cells isolated from human kidney cortex can be grown in primary cultures. These KCL express HLA class I and epithelial cells markers. They are devoid of detectable contamination with leukocytes. Coculture of KCL with

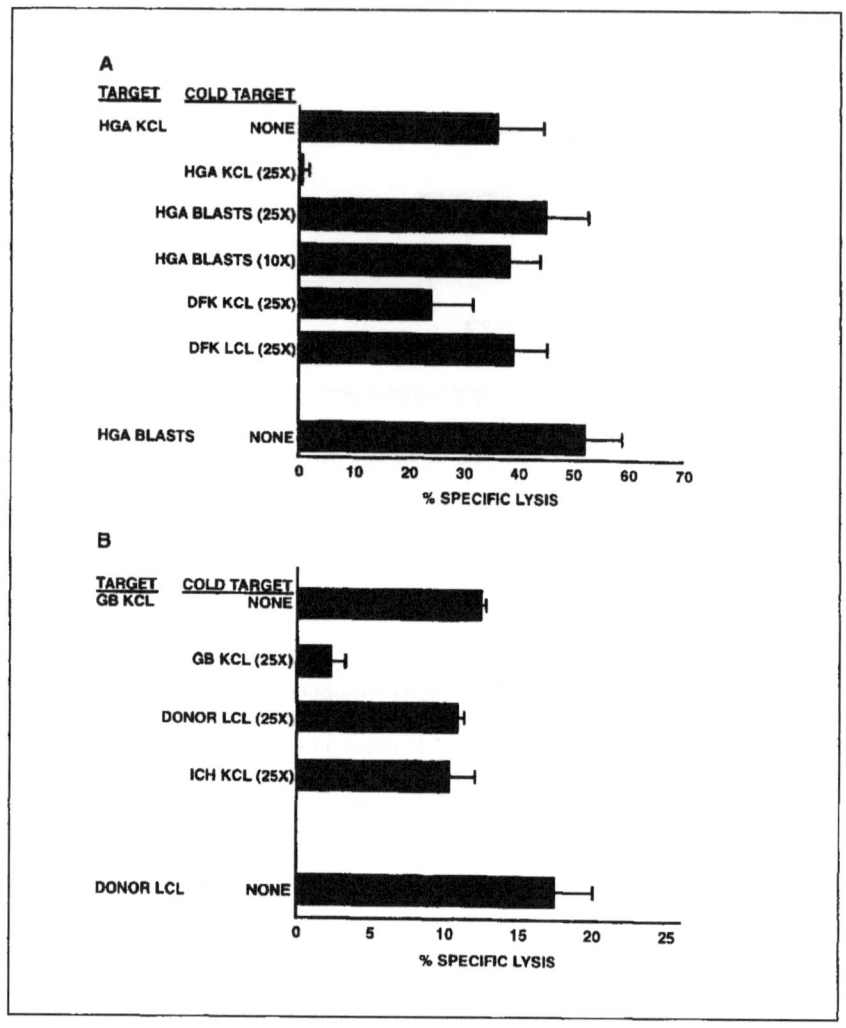

Fig. 16.2. Cold target inhibition analysis of GIL T-cell lines LMM (A) and GB (B) revealed there were kidney tissue-specific T cells within these bulk T-cell lines. GIL T cells were grown for 7 to 14 days in rIL-2 and then tested for lytic activity. Excess donor lymphoid targets (PHA stimulated donor lymph node cells, HGA blasts, or donor LCL) did not inhibit the lysis of donor KCL (HGA KCL and GB KCL).Using unlabeled excess donor KCL the lysis of ^{51}Cr-donor KCL was significantly inhibited. Third party targets, sharing no HLA antigens with donor (DFK KCL and ICH KCL) did not inhibit the lysis of donor KCL. Assays at an effector to target ratio of 10 to1.

normal peripheral blood leukocytes (PBL) from an HLA disparate individual results in a T-cell line that lyse the stimulating KCL in an HLA specific manner. Antibody blocking studies demonstrated that these T cells are primarily CD8[+] recognizing HLA class I alloantigens on KCL. When this T-cell line was cloned by limiting dilution we isolated some CD8[+] CTL that lysed the stimulator KCL but not an HLA identical LCL.[24] One of these kidney specific clones, DBS 1.5 was HLA-A3 restricted clone, and was further characterized with respect to tissue-specificity. HLA-A3 was transfected into a series of different tumor cell lines of diverse tissue origins. The transfected cells were used as targets for the tissue-specific clone. DBS 1.5 only lysed HLA-A3 transfected kidney epithelial cells.[25] These studies confirmed that the peptide recognized by this clone was found in kidney and was presented by the HLA-A3 molecule.

Antigenic peptides bound to MHC class I as well as class II molecules can be isolated following biochemical procedures originally described by Falk et al.[26] Using this approach we have isolated the peptide recognized by DBS 1.5.[25] It is a nine amino acid peptide having the sequence GPPGVTIVK. It conforms to the HLA-A3 sequence motif[27] containing nine amino acids with a lysine at P9 and is recognized by a kidney specific HLA A3 restricted CTL clone DBS 1.5.[25]

One of the difficulties in the identification of tissue-specific antigenic peptides is the requirement for tissues that express the same restricting HLA allele as that of the CTL as evidenced in our study.[25] To overcome this limitation, one can adapt an alternative strategy for isolation of HLA bound peptides that does not require a tissue expressing a specific HLA allele. This strategy involves biochemical isolation of three endoplasmic reticulum resident proteins namely heat shock protein gp96, protein disulfide isomerase, and calreticulin. All of these proteins have been shown to bind peptides that are translocated by transporter associated with antigen processing (TAP)[28] and thus serve as reservoirs of peptides prior to their loading on to MHC class I molecules. The repertoire of peptides that are bound to these proteins should represent the whole array of proteins inside the cell and should not be influenced by the MHC expression.

Conventional and Tissue-Specific Minor H Antigens

Clinical evidence for the important role minor H antigens play in transplantation comes mainly from studies involving HLA identical BMT recipients. In addition, renal allograft rejection occurs between HLA identical siblings and unrelated HLA identical renal transplants. Between October 1987 and December 1994 approximately 3300 zero HLA-mismatched cadaveric renal transplants were performed with 1 and 5 year survival rates of 86.9% and 70%, respectively.[11] These HLA matched transplants still require immunosuppression to prevent rejection, strongly suggesting a role for minor H antigens in renal allograft rejection.

Tissue-specific T-cells that recognize kidney specific minor H antigens have been described. Yamoto et al[29] isolated CTL clones specific for human minor H antigens from a patient who had been transplanted with the kidneys from his mother and two HLA identical sisters. These clones were MHC restricted by HLA-B35 expressed by both recipient and donors. They showed that the human minor H antigens recognized by these clones were found on kidney cells and their expression was increased when MHC class I was induced with IFN-γ. The immunogenic human minor H peptides were shown through bioassays to be contained in an HPLC fraction of a low molecular weight acid extracts of IFN-γ induced cultured kidney cells.[30]

Most of the tissue-specific clones we have isolated from graft infiltrating lymphocytes of acutely rejecting kidney allografts were specific for allo-MHC antigens (Fig. 16.1). However, one of these GIL isolated from a renal allograft rejection biopsy killed targets in a self MHC-restricted manner. Like Yamoto's T cells, these CTL demonstrated specific lytic activity against an HLA-B35 restricted minor H antigen, but they were tissue-specific recognizing HLA-B35 only on KCL and not LCL targets.[31]

Tissue-Specific Autoantigens and Autoimmunity

The concept of tissue-specific antigens and their possible role in various autoimmune diseases were postulated as early as 1960. Immunization with tissues such as thyroid, brain and testis have demonstrated the existence of organ specific antigens which resulted in the experimental production of autoimmune diseases. Pancreas specific alloantigens have also been reported in rabbits, monkeys, and in man[32] using serological methods.

There is increasing evidence in human autoimmune renal disease and animal models of autoimmune disease that T cells mediate tissue destruction in a tissue-specific manner. The role of T cells has been extensively studied in an animal model of anti-tubular basement membrane associated interstitial nephritis (anti-TBM disease).[33] The disease is induced in susceptible strains of mice by immunization with renal tubular antigens. Pathogenic CD4[+] and CD8[+] T-cell clones have been isolated. The latter have been shown to recognize renal tubular epithelial peptides in the context of MHC class I molecules.[33] Most clinically recognized human acute interstitial nephritis associated with systemic autoimmune disease occurs as either a primary process or secondary to underlying glomerular disease. Cell mediated immunity has been shown to be the primary effector mechanism in the disease, although anti-TBM antibodies and immune complexes have also been observed.[34] The antigens recognized by human renal autoreactive T cells are still unknown. In an attempt to uncover renal epithelial cell specific responses, we examined peripheral blood of 26 patients with renal autoimmune disease, with evidence of ANCA (anti-nuclear antibodies), IgA nephropathy, glomerulonephritis, and post-streptococcal nephritis, for cytolytic activity against KCL expressing autologous HLA class I antigens. Patient PBL were cultured for 7 days with a mixture of three KCL expressing HLA class I antigens of which some were matched and others mismatched with those of the patient. T cells were then tested in ^{51}Cr release assays using each of the three KCL and K562 (natural killer cell susceptible) as targets. To take into account the high level of NK activity in these cultures, K562 cells were also used as a cold targets, added to the CTL assays at concentrations equaling 25 times the number of ^{51}Cr-labeled KCL. 46% (12 of 26) of the patients showed high levels of CTL activity against HLA class I matched KCL at an E:T of 20:1. This lysis was specific in 42% (5 of 12) of these patients since cold target inhibition with K562 did not decrease the level of KCL lysis. These data strongly suggest that there is tissue-specific T-cell reactivity in the peripheral blood of patients with renal autoimmune disease. If renal tissue damage in autoimmune disease is due to recognition by T cells of immunogenic peptides presented by MHC class I antigens then it becomes important to identify the peptides epitopes. Their isolation and characterization will allow us to determine whether there are immunodominant self peptides being targeted by T cells.

Conclusion

The target of the T-cell mediated immune response to allografts is the MHC-peptide complex on the surface of the transplanted organ. Identification of these peptides may have important implications in how we determine what organs are suitable for transplantation. For example, identifying minor H antigens displayed on sibling cells may allow us to predict the most successful transplant donor among a family of possible donors. The contribution of tissue-specific minor H antigens and MHC alloantigens to the allograft response and their role in solid organ rejection is still controversial. It is possible that T cells directed against these antigens may play a prominent role in chronic rejection. Work in this area is ongoing.

The ultimate goal of transplantation research is to develop immunotherapeutic strategies that will induce organ specific tolerance in the recipient. Two approaches have been employed to induce tolerance to organ transplants. The first is to target the MHC-peptide complex recognized by the T cell. Accumulating evidence suggests the ability of both polymorphic and nonpolymorphic peptides of the MHC molecule to suppress the alloimmune response by antigen-specific and non-specific routes. A similar approach could be employed using tissue-specific peptide epitopes. This could be most important in those transplants involving six antigen matches or completely HLA-mismatched transplants if the tissue-specific peptides prove to be immunodominant. Identification of minor H peptides involved in GVHD in bone marrow transplantation where donor and recipient were genotypically identical has demonstrated that there are immunodominant minor H peptides involved in T cell-mediated GVHD.[35] If immunodominance is also found in solid organ transplant then the relevant peptides could be used therapeutically to induce antigen-specific suppression. Further, if allelic differences are noted between tissue-specific antigens minor H antigens, it may be possible to match for these antigenic variations especially with living related renal transplant donors.

The second approach is to target the TCR itself. Goss et al[36] have shown in a mouse transplant model that antibody depletion of $V\beta8^+$ T cells will significantly extend the survival of L^d vascularized heart and skin transplants in L^d-mutant BALB/C-$H2^{dm2}$ mice. A similar approach has been successful in treatment of experimental allergic encephalitis (EAE) where the disease has been treated with TCR peptides from the predominantly selected variable region.[37] This type of therapy is most effective in the initial stages of EAE, and likewise the early period post-transplant, when restricted $V\beta$ usage has been identified.[36] In human kidney the initial rejection episodes demonstrate a preferential TCR.[38] Therefore, targeting of tissue-specific T cells through their TCR with either clonotypic antibody or TCR peptides may block the initial rejection and decrease the need for aggressive conventional immunosuppression. These therapies will only be possible after the identification and characterization of the immunodominant minor and tissue-specific antigenic peptides.

Acknowledgment

This work was supported by NIH grant DK32253 (TM).

References

1. Hart DN, Fabre JW. Kidney-specific alloantigen system in the rat. Characterization and role in transplantation. J Exp Med 1980; 151:651-666.
2. Perkins HA, Gantan Z, Siegel S et al. Reactions of kidney cells with cytotoxic antisera: Possible evidence of kidney-specific antigens. Tissue Antigen 1975; 5:89-98.
3. Clayman MD, Michand L, Neilson EG. Murine interstitial nephritis. VI. Characterization of the B-cell response in anti-tubular basement membrane disease. J Immunol 1987; 139:2242-2249.
4. Wilson CB. Individual and strain differences in renal basement membrane antigens. Transplant Proc 1980; 12:69-73.
5. Mashimo S, Sakai A, Ochiai T, et al. The mixed kidney cell—lymphocyte reaction in rats. Tissue Antigens Copenhagen:Munksgaard, 1976:291-300.
6. Vegt PA, Buurman WA, vanderLinden CJ et al. Cell-mediated cytotoxicity toward canine kidney epithelial cells. Transplantation 1982; 33:465-469.
7. Esquenazi V, Fuller L, Pardo V et al. In vivo and in vitro induction of class II molecules on canine renal cells and their effect on the mixed lymphocyte kidney cell culture. Transplantation 1987; 44:680-692.
8. Manca F, Barocci S, Kunkl A et al. Recognition of donor fibroblast antigens by lymphocytes homing in the human grafted kidney. Transplantation 1983; 36:670-674.
9. Roth D, Fuller L, Esquenazi V et al. The biologic significance of the mixed lymphocyte kidney culture in humans. Transplantation 1985; 40:376-383.
10. Roth D, Flaa C, Fuller L et al. T cell lines and clones preferentially recognizing kidney-associated antigens in end-stage renal disease. Transplantation 1985; 40:686-693.
11. Smith CM, Ellison MD. UNOS. 1996 Annual Report-The US scientific registry of transplant recipients and the organ procurement and transplantation network. Bethesda, Maryland: U.S. Department of Health and Human Services, 1996.
12. Tesi RJ, Henry ML, Elkhammas EA et al. The frequency of rejection episodes after combined kidney-pancreas transplant—the impact on graft survival. Transplantation 1994; 58:424-430.

13. Calhoun R, Mohanakumar T, Flye MW. Tissue- and organ-specific immune responses: Role in human disease and allograft immunity. Transplant Rev 1996; 10:34-45.
14. Griffith BP, Hardesty RL, Trento A et al. Asynchronous rejection of heart and lungs following cardiopulmonary transplantation. Ann Thorac Surg 1985; 40:488-493.
15. Nakafusa Y, Goss JA, Mohanakumar T et al. Induction of donor-specific tolerance to cardiac but not skin or renal allografts by intrathymic injection of splenocyte alloantigen. Transplantation 1993; 55:877-882.
16. Joyce S, Flye MW, Mohanakumar T. Characterization of kidney cell-specific, non-MHC alloantigen using antibodies eluted from rejected human renal allografts. Transplantation 1988; 46:362-369.
17. Kirk AD, Ibrahim MA, Bollinger RR et al. Renal allograft-infiltrating lymphocytes: A prospective analysis of in vitro growth characteristics and clinical relevance. Transplantation 1992; 53:329-338.
18. Poindexter NJ, Steward NS, Shenoy S et al. Cytolytic T lymphocytes from renal allograft biopsies. Hum Immunol 1995; 44:43-49.
19. Poindexter NJ, Steward NS, Shenoy S et al. Renal allograft infiltrating lymphocytes: Frequency of tissue-specific lymphocytes. Hum Immunol 1997; 55:140-147.
20. Yard BA, Kooymans-Couthino M, Reterink T et al. Analysis of T-cell lines from rejecting renal allografts. Kidney Int 1993; 43:S133-S138.
21. Deckers JGM, Boonstra JG, Van der Kooij SW et al. Tissue-specific characteristics of cytotoxic graft-infiltrating T cells during renal allograft rejection. Transplantation 1997; 64:178-181.
22. Jutte NHPM, Knoop C, Heijse P et al. Human heart endothelial-cell-restricted allorecognition. Transplantation 1996; 62:403-406.
23. Hadley GA, Linders B, Mohanakumar T. Immunogenicity of MHC class I alloantigens expressed on parenchymal cells in the human kidney. Transplantation 1992; 54:537-542.
24. Hadley GA, Linders B, Mohanakumar T. Kidney cell-restricted recognition of MHC class I alloantigens by human cytolytic T-cell clones. Transplantation 1993; 55:400-404.
25. Poindexter NJ, Naziruddin B, McCourt DW et al. Isolation of a kidney-specific peptide recognized by alloreactive HLA-A3-restricted human CTL. J Immunol 1995; 154:3880-3887.
26. Falk K, Rotzschke O, Rammensee H. Cellular peptide composition governed by major histocompatibility complex class I molecules. Nature 1990; 348:248-251.
27. DiBrino M, Parker KC, Shiloach J et al. Endogenous peptides bound to HLA-A3 posses a specific combination of anchor residues that permit identification of potential antigenic peptides. Proc Natl Acad Sci USA 1993; 90:1508-1512.
28. Srivastava PK, Udono H, Blachere NE et al. Heat shock proteins transfer peptides during antigen processing and CTL priming. Immunogenetics 1994; 39:93-98.
29. Yamamoto H, Kariyone A, Akiyama N et al. Presentation of human minor histocompatibility antigens by HLA-B35 and HLA-B38 molecules. Proc Natl Acad Sci USA 1990; 87:2583-2587.
30. Beck Y, Sekimata M, Nakayama S et al. Expression of human minor histocompatibility antigen on cultured kidney cells. Eur J Immunol 1993; 23:467-472.
31. Poindexter NJ, Shenoy S, Howard T et al. Allograft infiltrating cytotoxic T lymphocytes recognize kidney specific human minor histocompatibility antigens. Clin Transpl 1996; 11:174-177.
32. Metzgar RS. Pancreas-specific alloantigens. Transplant Proc 1980; 12:123-128.
33. Meyers C and Kelly C. Effector mechanisms in organ-specific utoimmunity. I. Characterization of a CD8$^+$ T-cell line that mediates murine interstitial nephritis. J Clin Invest 1991; 88:408-416.
34. Kelly CG, Roth DA, Meyers CM. Immune recognition and response to the renal interstitium. Kidney Int 1991; 31:518-530.
35. denHaan JMM, Sherman NE, Blokland E et al. Identification of a graft-versus-host-disease associated human minor histocompatibility antigen. Science 1995; 268:1476-1479.

36. Goss JA, Pyo R, Flye MW et al. Major histocompatibility complex-specific prolongation of murine skin and cardiac allograft survival after in vivo depletion of Vβ+ T cells. J Exp Med 1993; 177:35-44.

37. Howell MD, Winters ST et al. Vaccination against experimental allergic encephalomyelitis with T-cell receptor peptides. Science 1989; 2:668-670.

38. Miceli MC, Finn OJ. T cell receptor β-chain selection in human allograft rejection. J Immunol 1989; 142:81-86.

CHAPTER 17

Human Tumor Antigens Recognized by Cytolytic T Lymphocytes: Towards Vaccination

Pierre van der Bruggen

Cytolytic T lymphocytes (CTLs) have been shown to mediate tumor rejection in several animal models. In humans, tumor immunologists are working under the assumption that T lymphocytes might be able to eradicate cancer cells as effectively as they do with virus-infected cells. An exciting challenge is to identify on cancer cells, specific antigens targeted by CTLs, and then to manipulate these antigens so that they become capable of initiating or amplifying a patient's native immune response, which would otherwise be insufficient.

Isolation of T Cells Recognizing Human Tumor Antigens

A large number of studies have shown that by cultivating irradiated tumor cells with blood lymphocytes of the same patient, i.e. autologous lymphocytes, it is possible to obtain responder cell populations that display a cytolytic activity against the tumor cells.[1] Such cytotoxic responder cells can also be generated from lymph node cells or tumor-infiltrating lymphocytes (TILs). These responder populations are then a source of tumor-specific CTL clones (Fig. 17.1). Of the human anti-tumor CTLs obtained so far, most have been generated against melanomas, mainly because metastatic melanoma cells are relatively easy to adapt to culture, providing cell lines that are convenient stimulator and target cells for the CTLs. Nevertheless, a number of tumors of other histological types have also been found to express antigens recognized by CTLs. An antigen recognized by a CD8[+] CTL consists of a complex between a peptide and an MHC class I molecule: HLA-A, B, or C. The antigenic peptides are produced inside the cell, mostly through degradation of cellular proteins. They are conveyed by specialized transporters from the cytosol into the endoplasmic reticulum, where some of them fit into a groove present on the surface of the HLA molecules. Once they have associated with the invariant β2-microglobulin chain, the MHC-peptide complexes move to the cell surface, where they become anchored in the cell membrane. Thus, peptides from endogenous proteins are continuously displayed and are ready to be scrutinized by CTLs.

Strategies for the Identification of Human Tumor Antigens

The characterization of a potential target antigen on tumor cells is the same as that for a minor histocompatibility (H) antigen: both require the identification of the antigenic peptide and its presenting HLA class I molecule. Therefore, the techniques that were used to

Minor Histocompatibility Antigens: From the Laboratory to the Clinic, edited by Derry Roopenian and Elizabeth Simpson. ©2000 Landes Bioscience.

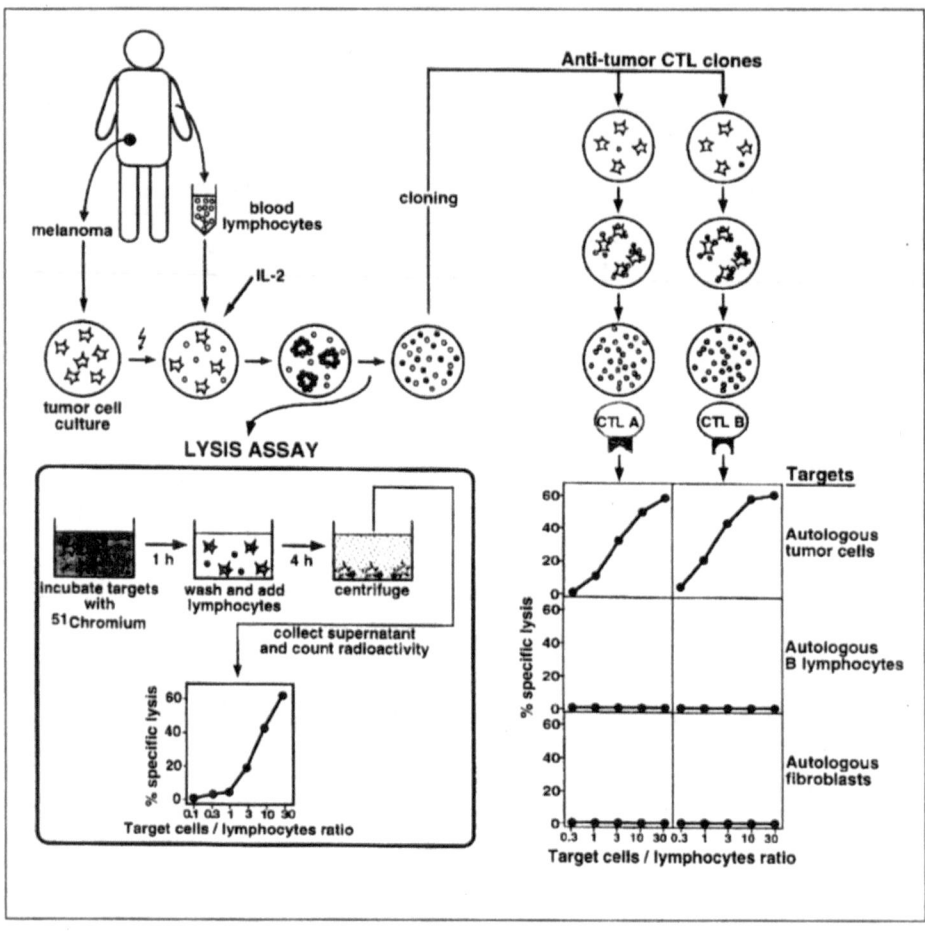

Fig. 17.1. How autologous anti-tumor cytolytic T-cell clones are obtained from the blood of melanoma patients. T cells isolated from the blood are stimulated in vitro with irradiated autologous tumor cells, in the presence of IL-2. The responder T cells can be cloned by limiting dilution. These T-cell clones are amplified and their lytic activity tested. A lysis assay, which can be conducted with the bulk anti-tumor T-cell population or with the T-cell clones, consists of mixing different numbers of T cells (from 30,000 down to 100/well) with a constant number (1,000 cells/well) of ^{51}Cr-labelled target cells. T cells and targets are incubated over 4 hours, during which the target cells are lysed by the CTLs.

identify tumor antigens may also be applied to the identification of minor H antigens. There are basically three methods that have been used to identify the peptides presented to tumor-specific T cells. The first is a genetic approach based on the transfection of recombinant DNA libraries into cells expressing the MHC presenting molecule, in order to isolate the gene encoding the protein from which the peptide is derived (Fig. 17.2).[2] Once the gene has been isolated, the antigenic peptide is deduced from the sequence of the putative protein. The second method is a biochemical purification of peptides eluted from the MHC class I molecules of the tumor cells. Tumor peptides are fractionated by HPLC, and the different fractions are then tested for their ability to sensitize target cells for lysis by relevant CTLs.

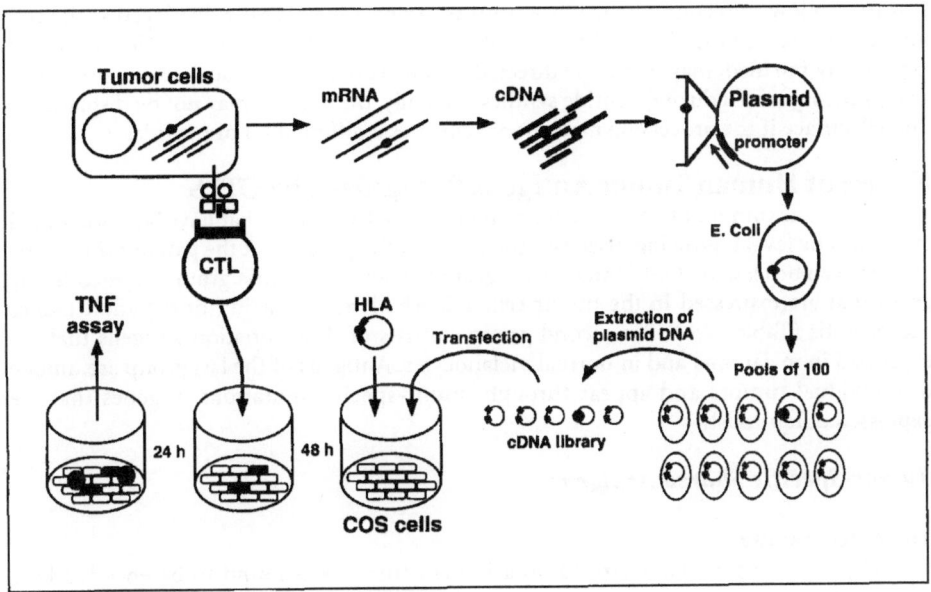

Fig. 17.2. The genetic approach used to identify tumor antigens recognized by CTLs. mRNA is extracted from cultured tumor cells. Single-stranded mRNA molecules are converted into double-stranded complementary DNA, using reverse transcriptase. Each cDNA molecule is inserted into a plasmid vector, which allows the amplification of the cDNA in bacteria. The type of plasmid used here also contains a eukaryotic promoter: when introduced into eukaryotic cells, simian COS cells in this case, the promoter ensures the transcription of the cDNA cloned in the plasmid. The plasmids are introduced into *E. coli*. The collection of transformed bacteria, each containing a plasmid with a cDNA fragment, is divided into pools of ± 100 bacteria, and these pools are amplified. The plasmid DNA is extracted from each pool of bacteria and transfected into COS cells. Plasmid DNA containing the gene encoding the appropriate HLA molecule is transfected together with the cDNA library. Each pool of 100 plasmids is transfected into a separate microculture of COS cells. Several hundred different pools are transfected. To detect the presence of COS cells expressing the antigen in one of the transfected populations, the CTL clone is added to all of them. The CTL recognizes its antigen and secretes TNF that is measured in the supernatant of the coculture medium.

The positive fraction is then further purified and sequenced. Sophisticated mass spectrometry methods are usually required to perform the peptide sequencing.[3] If post-translational modification of the protein occurs, the sequence of the peptide can not be deduced from the sequence of the protein and a biochemical approach is then the only way to identify the natural epitope presented at the cell surface.[4] The third method of identification is the reverse of the first two and has been used to identify antigens encoded by proteins known to be overexpressed or mutated in tumor cells. Candidate peptides carrying consensus anchor motives for a certain HLA are located, tested for HLA binding, and loaded on antigen-presenting cells that are then used to stimulate lymphocytes in vitro. [5,6] Despite being greeted with enthusiasm several years ago, this straightforward approach has not yet proved to be very productive for human tumor antigens. The major drawback appears to be that many peptide-specific CTLs do not recognize HLA-matched tumor cells expressing the protein endogenously. There are at least three different explanations for this paradox. In some cases, the CTLs activated in vitro with a high concentration of peptide may have an affinity too low to recognize the low amounts of peptide naturally displayed at the cell surface.[7] Other CTLs, obtained by stimulation with synthetic peptides that were not extensively purified, may in fact be directed against contaminants, such as the protective groups incorporated during peptide synthesis. Finally, the peptide may not be displayed at the cell surface if the processing machinery generates a different set of peptides.

Nature of Human Tumor Antigens Recognized by CTLS

The list of tumor antigens that are recognized by T cells and that have been identified at the genetic level is growing. Based on the pattern of expression of the parent protein, the antigens can be classified into three major groups. Those of the first group are encoded by genes that are expressed in the tumor cells but which are silent in normal adult tissues except testis (Table 17.1). The second group consists of differentiation antigens that are expressed in melanoma and in normal melanocytes. Antigens of the last group are unique to individual tumors and appear through tumor-specific mutations in genes that are expressed ubiquitously.

Tumor-Specific Shared Antigens

The *MAGE* Family

The first antigen characterized on a human tumor was found to be encoded by a previously unknown gene, designated *MAGE-1* (for Melanoma AntiGEn).[8] *MAGE-1* belongs to a family of at least 12 genes located on the long arm of chromosome X, several members of which are expressed in a significant proportion of tumors of different histological origins (Table 17.2).[9,10] Conservation of the main hydrophobic regions throughout the *MAGE* family suggests that the as yet unknown functions of all of the corresponding proteins might be related. In contrast, the promoters of the 12 *MAGE* genes show great variability, suggesting that a similar function could be regulated under different transcriptional controls, thus occurring at specific times and locations. Four additional *MAGE*-related genes are also located in the short arm of the same chromosome X.[11-13]

The only normal cells in which significant expression of *MAGE* genes has been detected are placental trophoblast and testicular germ cells.[14] Such antigens can be considered strictly tumor-specific, however, because these normal cells do not express MHC class I molecules and gene expression should not result in antigen presentation.[15] These conclusions are further strengthened by immunization studies carried out with the tumor antigen encoded by mouse gene P1A, which is also expressed in testis. Immunized male mice produced strong *P1A*-specific CTL responses but did not suffer testis inflammation nor any effect on fertility.[16]

Several *MAGE*-encoded peptides that bind to different HLA class I molecules and are recognized by tumor-specific CTLs have been identified (Table 17.1). Remarkably, all the relevant CTLs that recognize antigens encoded by *MAGE* genes were derived from the same melanoma patient, who had a metastasis but who enjoyed an unusually favorable clinical course. Several other patients expressing some or all of these genes were tested and no such CTLs were obtained by autologous tumor cell stimulation of their lymphocytes.

The *BAGE* and *GAGE* Families

Two other tumor antigens expressed by the melanoma cell line from which gene *MAGE-1* was cloned were found to be encoded by previously unknown genes, which were named *BAGE*[17] and *GAGE*[18] (Table 17. 1). The histological distribution of *BAGE* or *GAGE*-positive tumors is close to that of *MAGE*-positive tumors (Table 17.2).

Table 17.1. Tumor-specific antigens shared by different human tumors

Gene	Normal	MHC	Peptide	Position	Reference
MAGE1	testis	HLA-A1	EADPTGHSY	161-169	44
		HLA-Cw16	SAYGEPRKL	230-238	45
MAGE3	testis	HLA-A1	EVDPIGHLY	168-176	46
		HLA-A2	FLWGPRALV	271-279	47
		HLA-A24	IMPKAGLLI	195-203	48
		HLA-B44	MEVDPIGHLY	167-176	49
BAGE	testis	HLA-Cw16	AARAVFLAL	2-10	17
GAGE1/2	testis	HLA-Cw6	YRPRPRRY	9-16	18
RAGE1	retina	HLA-B7	SPSSNRIRNT	11-20	19
GnTV*	none	HLA-A2	VLPDVFIRC	38-64	21

*Aberrant transcript of N-acetyl glucosaminyl transferase V (*GnTV*) that is found only in melanomas (Table 17.2).

Table 17.2. Expression in tumor samples of genes encoding tumor-specific shared antigens*

Histological type	Percentage of tumors expressing:					
	MAGE-1	MAGE-3	BAGE	GAGE-1,2	RAGE-1	GnTV
Melanomas						
primary lesions	16	36	8	13	2	48
metastases	48	76	26	28	5	
Non-Small Cell Lung Carcinomas	49	47	4	19	0	0
Head and Neck tumors	28	49	8	19	2	0
Bladder carcinomas	22	36	15	12	5	0
Sarcomas	14	24	6	25	14	0
Mammary carcinomas	18	11	10	9	1	0
Prostatic carcinomas	15	15	0	10	0	0
Colorectal carcinomas	2	17	0	0	0	0
Renal cell carcinomas	0	0	0	0	2	0
Leukemias and lymphomas	0	0	0	1	0	0

*Expression was measured by RT-PCR on total RNA using primers specific for each gene.

The *RAGE* Gene

Most antigens of this group have been characterized on melanoma, but a number of tumors of other histological types have also been found to express such antigens. *RAGE* was recently found to code for an antigen recognized by CTLs on a human renal cell carcinoma (Table 17.1).[19] Like the *MAGE* gene, *RAGE* is not expressed in most normal tissues, but is activated in a number of tumors of various histological types (Table 17.2). Interestingly, it is also expressed in normal retina. However, most intra-ocular tissues, including retina, do not appear to express MHC class I molecules.[20] Moreover, the eye is an immune privileged site and contains a variety of potent immunosuppressive agents that protect it from destructive inflammatory reactions.

N-acetylglucosaminyltransferase V

Recently, a new mode of origin for shared tumor-specific antigens was discovered.[21] Here, it seems that *N-acetyl-glucosaminyltransferase V* (*GnTV*, Table 17.1), a gene that is ubiquitously expressed, contains an intron which in turn appears to carry near its end a promoter that is activated only in melanoma cells. This atypical activation, which occurs in more than 50% of melanomas, results in a message containing a new open reading frame, which codes for the antigenic peptide in its intronic part.

Melanocytic Differentiation Antigens

Surprisingly, a large number of CTL clones generated against autologous melanomas and recognizing tumor antigens presented by HLA-A2 molecules were found also to be able to recognize HLA-A2-positive normal melanocytes. This observation indicates that the corresponding antigens are actually melanocytic differentiation antigens, which are also expressed by melanomas. Four genes encoding such melanocytic differentiation antigens have been identified: tyrosinase, Melan-A / MART-1, gp100 and gp75.[22-28] Tyrosinase is an enzyme involved in the synthesis of melanin, whereas the role of the three other proteins is unknown. The four corresponding genes are expressed in no other normal tissues except melanocytes, but they are expressed by virtually all melanoma samples. Most of the identified antigenic peptides encoded by these genes are presented by HLA-A2, although other HLA-peptide combinations have been found. The pattern of CTL precursors directed against these differentiation antigens appears to be very different from that observed with the *MAGE*-like antigens. In the former, most melanoma patients have CTL precursors that can be readily restimulated in vitro with autologous tumor cells.[29,30] TIL populations also contain these CTLs.[31] But what of the potential side-effects of active or passive immunization against melanoma differentiation antigens? Concerns have been expressed on this account, not so much for the skin, where an acceptable occurrence might be vitiligo due to the destruction of melanocytes, but rather for the retina, where melanocytes are present in the choroid layer. However, vitiligo has been associated with good prognoses in melanoma and also with adoptive transfer of TILs, without noticeable eye lesions.[32,33] Carefully devised immunotherapy trials based on these antigens therefore seem justifiable. The role of CTLs against such antigens in melanoma rejection is not clear, but is supported by the reported association of vitiligo with prolonged survival and spontaneous regression of melanoma.[34] Nevertheless, the potential ophthalmic toxicity of active immunotherapy needs to be evaluated very carefully.

Antigens Resulting from Mutations

Initial work carried out on mouse tumors has revealed that point mutations also generate antigens recognized on tumors by autologous CTLs.[35,36] The mutations are located in the region coding for the antigenic peptide and each changes one amino acid of an antigenic peptide. This can have one of two possible consequences: The modified amino acid either enabled the peptide to bind to a class I molecule (new agretope) or constituted a new epitope recognized by a CTL. A mutation in virtually any gene can conceivably result in the appearance of a new antigen on a cell. Accordingly, an infinite variety of antigens can be produced by random mutations. The great majority of mouse tumor antigens identified to date are the result of mutations, which may be because many of the mouse tumors studied were induced with radiation or other carcinogens. However, antigens of this category have also been found on spontaneous mouse tumors and on human tumors. Two such mutations affecting human genes CDK4 and β-catenin may be involved in oncogenesis, since they were found in several independent tumors and have a demonstrated effect on the activity of the encoded proteins.[37-39] The CDK4 mutation prevents the protein from binding to its inhibitor, p16. This appears to alter the regulation of the cell cycle, favoring uncontrolled growth of the tumor cells. The β-catenin mutation results in stabilization of the protein, which favors the constitutive formation of complexes with transcription factors, such as Lef-1. Constitutive β-catenin/Lef-1 complexes may result in persistent transactivation of as yet unidentified target genes, thereby stimulating cell proliferation or inhibiting apoptosis.[39,40]

Another mutation which may antagonize apoptosis was identified recently with CTLs specific for a human squamous-cell carcinoma.[41] The antigen is encoded by a mutated form of the *CASP-8* gene, coding for protease caspase-8. This protease is required for induction of apoptosis through the Fas and TNFR1 receptors, and the ability of the altered protein to trigger apoptosis appears to be reduced relative to the normal caspase-8. Combined with the observations made with the CDK4 and β-catenin mutations, this suggests that a fraction of point mutations generating tumor antigens also play a role in tumor transformation or progression.

It would seem reasonable to assume that point mutation-generated antigens are specific only for the tumor cells, and that the CTL precursors directed against these antigens will not have undergone any of the depletion or anergy that accompanies natural tolerance. However, the fact that they are expected to be unique for an individual tumor, or restricted to very few, currently makes them unwieldy for the development of therapeutic cancer vaccines.

Towards Vaccination

The identification of genes coding for human tumor antigens heralds a new era in the search for an effective, specific immunotherapy for cancer (Fig. 17.3). By understanding the molecular nature of these antigens, it is now possible to select the patients whose tumors actually express a given antigen. Eligible tumors would need to express the relevant gene and have the appropriate HLA class I allele. This can be tested readily by RT-PCR on RNA extracted from a small tumor sample. The fraction of tumors expressing a given antigen can be calculated from the frequency of expression of the relevant gene in that tumor type and from the frequency of the presenting class I molecule in the population. More than 60% of melanomas from Caucasian patients bear one of the presently defined antigens encoded by *MAGE, BAGE* and *GAGE*. For other cancers, such as head and neck tumors and bladder cancer, the frequencies range from 28% to 40%. Furthermore, the definition of the molecular nature of tumor antigens facilitates the rational design of specific vaccine preparations.

A vast range of procedures have been proposed for T-cell immunization. Inoculation with peptides or proteins mixed with adjuvants is one possibility. Alternatively, this could be done ex vivo: specialized antigen-presenting cells such as dendritic cells can be isolated from the blood, incubated in vitro with protein or peptide, and inoculated back into the patient. A second option is the use of recombinant constructs. Expression systems have been designed in several microorganisms for the purpose of vaccination: BCG, vaccinia, adenovirus and retrovirus. Recombinant viruses or bacteria can be administered to the

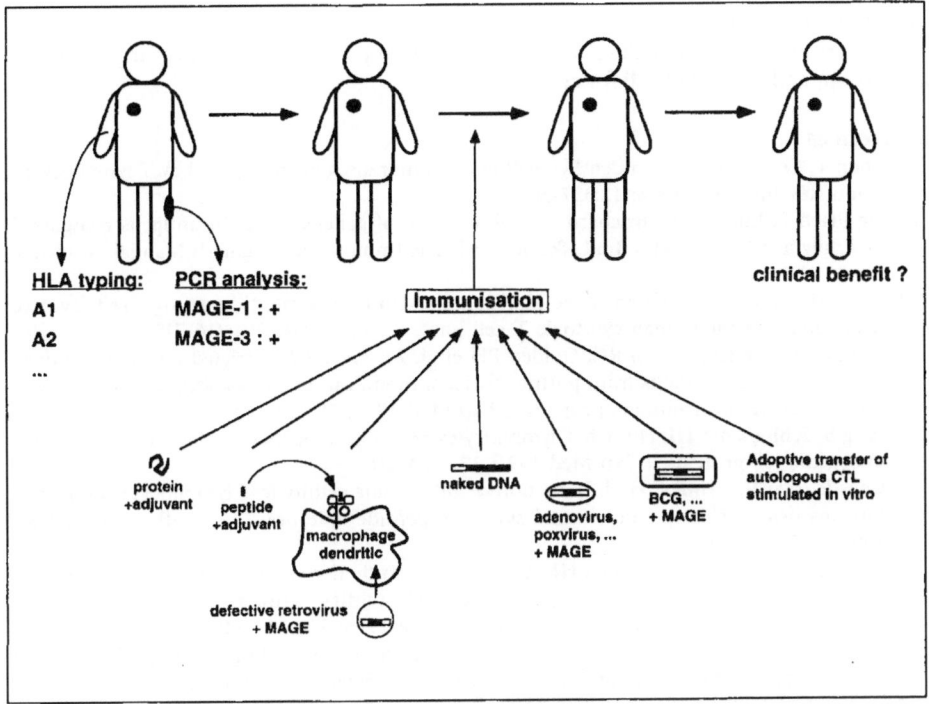

Fig. 17.3. Examples of specific immunotherapy of cancer using defined tumor antigens. Any cancer patient can be HLA-typed, and the expression of the different genes encoding tumor antigens can be tested by reverse transcription PCR on RNA extracted from a very small piece of tumor. The antigens that are expressed in the tumor can then be incorporated into vaccine preparations, in order to induce an anti-tumor immune response. Several modalities of immunization with an antigen are available, and it is presently not known which of them will prove to be the most efficient.

patients, or used in vitro to infect antigen-presenting cells isolated from the blood. Inoculation of DNA encoding the antigen is yet another possibility. Finally, if active immunization fails, an alternative approach is the adoptive transfer of autologous T lymphocytes stimulated in vitro with the antigen.

It is impossible now to predict the clinical outcome of such strategies, but preliminary results with a peptide encoded by gene *MAGE-3* are promising.[42] It is important to bear in mind, however, that the clinical efficacy of immunization may be limited by several factors. Antigen-loss variants may not present a major problem because most tumors seem to carry several different tumor antigens, and so the loss of one antigen may be circumvented by immunizing simultaneously against several antigens. A more serious problem, however, is the frequent occurrence of HLA-loss variants.[43]

It is our hope that responses will be obtained in some patients, and that the careful study of the lymphocytes and the tumor cells of these patients will produce not only a harvest of additional antigens but, more significantly, a better understanding of what constitutes an effective anti-tumor response.

Acknowledgments
We are grateful to Nathalie Krack and Simon Mapp for help in the preparation of the manuscript and to Uzi Gileadi for critical reading.

References
1. Boon T, Cerottini J-C, Van den Eynde B et al. Tumor antigens recognized by T lymphocytes. Annu Rev Immunol 1994; 12:337-365.
2. De Plaen E, Lurquin C, Brichard V et al. Cloning of genes coding for antigens recognized by cytolytic T lymphocytes. In: Lefkovits I, ed. The Immunology Methods Manual. Academic Press, 1997:692-718.
3. Cox AL, Skipper J, Chen Y et al. Identification of a peptide recognized by five melanoma-specific human cytotoxic T-cell lines. Science 1994; 264:716-719.
4. Skipper JCA, Hendrickson RC, Gulden PH et al. An HLA-A2-restricted tyrosinase antigen on melanoma cells results from posttranslational modification and suggests a novel pathway for processing of membrane proteins. J Exp Med 1996; 183:527-534.
5. Jung S, Schluesener HJ. Human T lymphocytes recognize a peptide of single point-mutated, oncogenic ras proteins. J Exp Med 1991; 173:273-276.
6. Celis E, Tsai V, Crimi C et al. Induction of anti-tumor cytotoxic T lymphocytes in normal humans using primary cultures and synthetic peptide epitopes. Proc Natl Acad Sci USA 1994; 91:2105-2109.
7. Dahl AM, Beverley PCL, Stauss HJ. A synthetic peptide derived from the tumor-associated protein mdm2 can stimulate autoreactive, high avidity cytotoxic T lymphocytes that recognize naturally processed protein. J Immunol 1996; 157:239-246.
8. van der Bruggen P, Traversari C, Chomez P et al. A gene encoding an antigen recognized by cytolytic T lymphocytes on a human melanoma. Science 1991; 254:1643-1647.
9. Rogner UC, Wilke K, Steck E et al. The melanoma antigen gene (*MAGE*) family is clustered in the chromosomal band Xq28. Genomics 1995; 29:725-731.
10. De Plaen E, Arden K, Traversari C et al. Structure, chromosomal localization and expression of twelve genes of the *MAGE* family. Immunogenetics 1994; 40:360-369.
11. Muscatelli F, Walker AP, De Plaen E et al. Isolation and characterization of a new *MAGE* gene family in the Xp21.3 region. Proc Natl Acad Sci USA 1995; 92:4987-4991.
12. Dabovic B, Zanaria E, Bardoni B et al. A family of rapidly evolving genes from the sex reversal critical region in Xp21. Mammal Genome 1995; 6:571-580.
13. Lurquin C, De Smet C, Brasseur F et al. Two members of the human *MAGE-B* gene family located in Xp21.3 are expressed in tumors of various histological origins. Genomics 1997; 46:397-408.
14. Takahashi K, Shichijo S, Noguchi M et al. Identification of *MAGE-1* and *MAGE-4* proteins in spermatogonia and primary spermatocytes of testis. Cancer Res 1995; 55:3478-3482.
15. Haas GG Jr, D'Cruz OJ, De Bault LE. Distribution of human leukocyte antigen-ABC and -D/DR antigens in the unfixed human testis. Am J Reprod Immunol Microbiol 1988; 18:47-51.
16. Uyttenhove C, Godfraind C, Lethé B et al. The expression of mouse gene *P1A* in testis does not prevent safe induction of cytolytic T cells against a *P1A*-encoded tumor antigen. Int J Cancer 1997; 70:349-356.
17. Boël P, Wildmann C, Sensi M-L et al. *BAGE*, a new gene encoding an antigen recognized on human melanomas by cytolytic T lymphocytes. Immunity 1995; 2:167-175.
18. Van den Eynde B, Peeters O, De Backer O et al. A new family of genes coding for an antigen recognized by autologous cytolytic T lymphocytes on a human melanoma. J Exp Med 1995; 182:689-698.
19. Gaugler B, Brouwenstijn N, Vantomme V et al. A new gene coding for an antigen recognized by autologous cytolytic T lymphocytes on a human renal carcinoma. Immunogenetics 1996; 44:323-330.
20. Abi-Hanna D, Wakefield D, Watkins S. HLA antigens in ocular tissues. I. In vivo expression in human eyes. Transplantation 1988; 45:610-613.

21. Guilloux Y, Lucas S, Brichard VG et al. A peptide recognized by human cytolytic T lymphocytes on HLA-A2 melanomas is encoded by an intron sequence of the N-acetylglucosaminyltransferase V gene. J Exp Med 1996; 183:1173-1183.

22. Brichard V, Van Pel A, Wölfel T et al. The tyrosinase gene codes for an antigen recognized by autologous cytolytic T lymphocytes on HLA-A2 melanomas. J Exp Med 1993; 178:489-495.

23. Wölfel T, Van Pel A, Brichard V et al. Two tyrosinase nonapeptides recognized on HLA-A2 melanomas by autologous cytolytic T lymphocytes. Eur J Immunol 1994; 24:759-764.

24. Coulie PG, Brichard V, Van Pel A et al. A new gene coding for a differentiation antigen recognized by autologous cytolytic T lymphocytes on HLA-A2 melanomas. J Exp Med 1994; 180:35-42.

25. Kawakami Y, Eliyahu S, Delgado CH et al. Identification of a human melanoma antigen recognized by tumor-infiltrating lymphocytes associated with in vivo tumor rejection. Proc Natl Acad Sci USA 1994; 91:6458-6462.

26. Bakker ABH, Schreurs MWJ, de Boer AJ et al. Melanocyte lineage-specific antigen gp100 is recognized by melanoma-derived tumor-infiltrating lymphocytes. J Exp Med 1994; 179:1005-1009.

27. Kawakami Y, Eliyahu S, Jennings C et al. Recognition of multiple epitopes in the human melanoma antigen gp100 by tumor-infiltrating T lymphocytes associated with in vivo tumor regression. J Immunol 1995; 154:3961-3968.

28. Wang R-F, Parkhurst MR, Kawakami Y et al. Utilization of an alternative open reading frame of a normal gene in generating a novel human cancer antigen. J Exp Med 1996; 183:1131-1140.

29. Brichard VG, Herman J, Van Pel A et al. A tyrosinase nonapeptide presented by HLA-B44 is recognized on a human melanoma by autologous cytolytic T lymphocytes. Eur J Immunol 1996; 26:224-230.

30. Sensi ML, Traversari C, Radrizzani M et al. Cytotoxic T-lymphocyte clones from different patients display limited T-cell-receptor variable-region gene usage in HLA-A2-restricted recognition of the melanoma antigen Melan-A/MART-1. Proc Natl Acad Sci USA 1995a; 92:5674-5678.

31. Robbins PF, El-Gamil M, Kawakami Y et al. Recognition of tyrosinase by tumor-infiltrating lymphocytes from a patient responding to immunotherapy. Cancer Res 1994; 54:3124-3126.

32. Bystryn J-C, Darrell R, Friedman RJ et al. Prognostic significance of hypopigmentation in malignant melanoma. Arch Dermatol 1987; 123:1053-1055.

33. Richards JM, Mehta N, Ramming K et al. Sequential chemoimmunotherapy in the treatment of metastatic melanoma. J Clin Oncol 1992; 10:1338-1343.

34. Rosenberg SA, White DE. Vitiligo in patients with melanoma: Normal tissue antigens can be target for cancer immunotherapy. J Immunother 1996; 19:81-84.

35. De Plaen E, Lurquin C, Van Pel A et al. Immunogenic (tum⁻) variants of mouse tumor P815: Cloning of the gene of tum⁻ antigen P91A and identification of the tum⁻ mutation. Proc Natl Acad Sci USA 1988; 85:2274-2278.

36. Lurquin C, Van Pel A, Mariamé B et al. Structure of the gene coding for tum- transplantation antigen P91A. A peptide encoded by the mutated exon is recognized with Ld by cytolytic T cells. Cell 1989; 58:293-303.

37. Wölfel T, Hauer M, Schneider J et al. A p16^{INK4a}-insensitive CDK4 mutant targeted by cytolytic T lymphocytes in a human melanoma. Science 1995; 269:1281-1284.

38. Robbins PF, El-Gamil M, Li YF et al. A mutated β-catenin gene encodes a melanoma-specific antigen recognized by tumor infiltrating lymphocytes. J Exp Med 1996; 183:1185-1192.

39. Rubinfeld B, Robbins P, El-Gamil M et al. Stabilization of β-catenin by genetic defects in melanoma cell lines. Science 1997; 275:1790-1792.

40. Peifer M. β-catenin as oncogene: The smoking gun. Science 1997; 275:1752-1753.

41. Mandruzzato S, Brasseur F, Andry G et al. A CASP-8 mutation recognized by cytolytic T lymphocytes on a human head and neck carcinoma. J Exp Med 1997; 186:785-793.

42. Marchand M, Weynants P, Rankin E et al. Tumor regression responses in melanoma patients treated with a peptide encoded by gene *MAGE-3*. Int J Cancer 1995; 63:883-885.

43. Garrido F, Ruiz-Cabello F, Cabrera T et al. Implications for immunosurveillance of altered HLA class I phenotypes in human tumours. Immunol Today 1997; 18:89-95.
44. Traversari C, van der Bruggen P, Luescher IF et al. A nonapeptide encoded by human gene *MAGE-1* is recognized on HLA-A1 by cytolytic T lymphocytes directed against tumor antigen MZ2-E. J Exp Med 1992; 176:1453-1457.
45. van der Bruggen P, Szikora J-P, Boël P et al. Autologous cytolytic T lymphocytes recognize a *MAGE-1* nonapeptide on melanomas expressing HLA-Cw*1601. Eur J Immunol 1994; 24:2134-2140.
46. Gaugler B, Van den Eynde B, van der Bruggen P et al. Human gene *MAGE-3* codes for an antigen recognized on a melanoma by autologous cytolytic T lymphocytes. J Exp Med 1994; 179:921-930.
47. van der Bruggen P, Bastin J, Gajewski T et al. A peptide encoded by human gene *MAGE-3* and presented by HLA-A2 induces cytolytic T lymphocytes that recognize tumor cells expressing *MAGE-3*. Eur J Immunol 1994; 24:3038-3043.
48. Tanaka F, Fujie T, Tahara K et al. Induction of antitumor cytotoxic T lymphocytes with a *MAGE-3*-encoded synthetic peptide presented by human leukocytes antigen-A24. Cancer Res 1997; 57:4465-4468.
49. Herman J, van der Bruggen P, Luescher I et al. A peptide encoded by human gene *MAGE-3* and presented by HLA-B44 induces cytolytic T lymphocytes that recognize tumor cells expressing *MAGE-3*. Immunogenetics 1996; 43:377-383.

Index

V

Y